PILOT'S HANDBOOK OF

Flight Operating Instructions

FOR

Models B-25C and B-25D

AIRPLANES POWERED WITH 2 MODEL R-2600-13 ENGINES

Manufactured by North American Aviation, Inc., Inglewood, California

Contract: W535 ac-16070, 23471 (B-25C)   *Spec. Type C-213, Model NA-954*
Contract: W535 ac-19341   (B-25D)   *Spec. Type C-213, Model NA-5106*

NOTE: *This Technical Order replaces T.O. No. 01-60GB-1 dated March 10, 1942.*

NOTICE: *This document contains information affecting the National Defense of the United States within the meaning of the Espionage Act (U.S.C. 50:31:32). The transmission of this document or the revelation of its contents in any manner to an unauthorized person is prohibited.*

PUBLISHED BY AUTHORITY OF THE COMMANDING GENERAL
ARMY AIR FORCES, BY THE CHIEF, FIELD SERVICES,
AIR SERVICE COMMAND, PATTERSON FIELD, FAIRFIELD, OHIO

November 1, 1942

AIRCRAFT MANUAL SERIES

Curtiss Standard JN4-D—Jenny
Ford Trimotor
Lockheed P-38 Lightning
Bell P-39 Airacobra
Curtiss P-40 Warhawk
Republic P-47 Thunderbolt
North American F-51 Mustang
Northrop P-61 Black Widow
Chance Vought F4U Corsair
Flying Wings of Northrop
Messerschmitt ME-262 Sturmvogel
DeHavilland Mosquito
Supermarine Spitfire
Hawker Hurricane
Boeing B-17 Flying Fortress
Bell P-63 Kingcobra
X-15 Research Airplane
F-82 Twin Mustang
J-3 Piper Cub
Story of the Texan AT-6
Lockhead F-80 Shooting Star
Grumman F6F Hellcat
B-24 Liberator
Guide to Pre-1930 Aircraft Engines
North American B-25 Mitchell Bomber
Grumman Wildcat
Curtiss OX-5 Aeronautical Engine

TECHNICAL & REFERENCE MANUALS

Helicopter Design and Data Manual
Aircraft Detail Design Manual 2nd Ed.
Aircraft Hardware Standards Manual
 and Engineering Reference
Handbook of Airfoil Sections for Light Aircraft
1929 Airline Schedule of Commercial Transport
World War II International Aircraft
 Recognition Manual

AVIATION
 PUBLICATIONS

217 E. WASHINGTON ST.
P.O. BOX 357
APPLETON
WISCONSIN 54912
U.S.A.

ISBN-0-87994-004-2

Litho in U.S.A. at Graphic Communications Center
with production by Badger Printing Division
Appleton, Wisconsin 54912

THE B-25 "MITCHELL" BOMBER STORY

by Leo J. Kohn

(All photos from Collect-Air Photos)

Assigned to Civil Air Patrol Headquarters, this former B-25J-25-NC, AAF 44-30120A, became a VB-25-1 fitted with curtains, the half-ring exhaust system, and a clear windshield.

INTRODUCTION

If it had no greater claim to fame than that it was the first airplane by which the Americans took the war back to the enemy in the famous Doolittle raid on Tokyo, it would still have been a great airplane. A docile, adaptable, operationally efficient airplane with excellent all-round performance and particularly good handling characteristics, the B-25 **Mitchell** was a favorite with almost every air crew assigned to it.

It was manufactured in greater quantities than any other American twin-engined bomber, with no less than 9816 being built. Most of these went to Allied air forces and the airplane was one of the most popular of the American-supplied types with Allied air crews.

There was no other medium-bomber type ever subjected to as many modifications, either at the factory or under primitive conditions in the field, as was the B-25. The simple but rugged structure and somewhat flat lines allowed a maximum of modifications to be made with a minimum of tools at advanced combat bases. Added to the airplane's in-

herent ease of operation, it provided the B-25 with greater versatility than any other single airplane in its class. Dollar for dollar, it returned more productive value than almost any other airplane.

The B-25 **Mitchell** served in all principal war zones. It played a sizable role in smashing Rommel's Afrika Korps, aided in the great Russian winter offensive, shot and bombed its way up through the Pacific, and also saw combat in Alaska, the Mediterranean, Europe, Burma, the American shorelines, and even later in Korea.

In a single action in the Pacific area, Fifth Air Force units led by Major Ed Larner attacked and completely destroyed in the Bismark Sea a total of ten warships and 12 transports or merchant vessels. In addition, 61 enemy aircraft were destroyed with an additional 25 probables, and ten damaged. Total Army Air Forces personnel losses were 12, while the Japanese lost 12,700 officers and men, the equivalent of a full division. To accomplish such feats, every B-25 in the Pacific area that could be acquired was hurriedly modified to perform one or another specific mission.

DEVELOPMENT OF THE DESIGN

The B-25 was a supreme achievement for the relatively young manufacturer known as North American Aviation, Inc. Prior to this, the company was best known for its design and manufacture of observation airplanes, and its line of basic and advanced trainers and various export versions of these trainer types. The company's experience with multi-engined bombers was negligible. In 1938 it built a single XB-21 medium bomber but the Army Air Corps showed no further interest in it.

In 1939 the Air Corps issued what was known as a "Circular Proposal" for a twin-engined attack bomber. Several manufacturers, including North American, responded. In its response to this Proposal, North American conceived an advanced project, followed by the construction of a prototype aircraft designated as the NA-40.

The ill-fated North American NA-40, X14221, was the company's entry in an attack-bomber competition. Its tricycle-landing-gear layout was its only contribution to the B-25 design.

The NA-40 Attack Bomber prototype was built with private capital, and was completed in January, 1939, and was flown for the first time later that same month by Paul Balfour. For powerplants it mounted two Pratt & Whitney 14-cylinder R-1830-S6C3G engines of 1100 hp each.

In February the engines were changed to 1350-hp Wright GR-2600-A71's, and the airplane now became known as the NA-40-2, or alternately the NA-40B.

In March, following preliminary company flight-testing, the airplane was flown from Los Angeles to Wright Field, Ohio for the Air Corps' Attack Bomber competition and evaluation. This was followed by two weeks of additional testing, at which time the NA-40-2 was taken up by Major Younger Pitt. During an attempt to enter the approach pattern to Wright Field, control was lost and the airplane crashed. Though the NA-40-2 was consumed by fire, the crew escaped unharmed, and the design and construction of the NA-40-2 was absolved of blame.

North American had great expectations for their NA-40 beyond Air Corps acceptance, and had ex-

port versions planned up through the NA-40-7. Meager reports out of the flight-test program suggested that the NA-40-2 was capable of outstanding performance. With the loss of the prototype and with time running out, the equally outstanding Douglas DB-7 was selected and ultimately became the A-20. With that, the company dropped any further development of the NA-40 series.

By the beginning of 1939, the threat and actual incidence of war was heating up in Europe. On January 25, 1939, the Army Air Corps sent out invitations to manufacturers to submit design data for a medium bombardment airplane to replace the already obsolete Douglas B-18 bombers. The manufacturers were given very little time to develop proposals.

North American required 40 days to work up its designs and data, and there was no time to build a scale model and conduct wind-tunnel tests. The design was designated the NA-62, and there was no prototype built. The NA-40-2 is generally conceded to have been the prototype of what was to become the B-25 series, but this is not so. It was merely a private venture that was unsuccessful. However, between the XB-21 and the NA-40-2, both twin-engined bombers, only the NA-40-2 had the tricycle landing gear, and this provided a good starting point for the medium bomber which was also to be so equipped according to the Air Corps specifications.

Under the direction of J. L. Atwood and R. H. Rice, North American Aviation submitted 78 different design combinations to the Army, some calling for Wright engines, others for Pratt & Whitney engines. Crew arrangements varied according to the designs, and provisions were made for a single as well as a twin rudder arrangement. Eventually, the twin-rudder layout was selected because it enabled the provision of greater fire power and field of fire to the rear of the bomber.

The NA-62 was much cleaner aerodynamically than the NA-40-2, having a different wing and new-type engine nacelles. The fuselage lines and structure were quite different, as was the tail structure.

North American's aerodynamics section was confronted with a tremendous job in making performance calculations from the various designs proposed to the Air Corps. Everything was on paper and there were no wind-tunnel-test data to guide them.

When on September 10, 1939 the Air Corps completed its study of the various proposals submitted by the aircraft manufacturers, the twin-tail NA-62 was one of two medium bombers selected for production. The NA-62 became the B-25, and the other medium bomber chosen was the Martin B-26. On September 20, 1939, North American Aviation received a contract for 184 B-25 bombers.

When word was received that the NA-62 design had been accepted, a one-ninth scale model was immediately built, and was one of the first wind-tunnel models to have electric motors to spin its propellers. This allowed the aerodynamicists to make more accurate performance calculations, and also aided in measuring of loading capacity and stalling characteristics. Careful calculations made at the California Institute of Technology in Pasadena paid dividends in succeeding months during Army Air Corps tests. The design of the B-25 required approximately 156,000 engineering man-hours in the production of 8500 original drawings.

A full-scale wooden mock-up of the B-25 was built complete with seats and controls and full instrumentation at the same time that the wind-tunnel model was being built. After only a few minor changes, an Army Mock-Up Board approved the layout on November 4, 1939.

The original design of the B-25, as accepted by the Army Air Corps, called for a high-speed mid-wing monoplane powered by two 1650-hp Wright **Cyclones** with superchargers, driving constant-speed Hamilton Standard **Hydromatic** three-bladed propellers of 12-ft. 7-in. diameter. The bomb load was calculated and a crew of five was planned — pilot, co-pilot, bombardier, radio-operator, and gunner. The bombardier and radio-operator would man the guns in their respective compartments. Access to the nose was via a tunnel under the pilot, and to the rear via a crawlway over the bomb-bay. A speed of over 300 m.p.h. was guaranteed. Armament was to consist of a flexible .30-cal. gun in the nose that could be used in three positions, and three .30-cal. waist guns, one on either side and one on top. A .50-cal. gun in the tail was to be operated by a gunner in a sitting position.

The first North American B-25, AAC 40-2165, with the pre-war stars and striped tail insignia, shows the constant wing dihedral. Except for that the basic lines and structure remained unchanged from the start.

So, no **XB-25** was ordered, and the B-25 went directly into production off the drawing boards. The first 24 B-25's were to be delivered in 1940, and the first of these was test-flown without incident on August 19, 1940 after a static-test airframe had been shipped to Wright Field the previous July 4. Succeeding test flights proved the performance calculations made by the engineers to be on the conservative side. Its flying speed was highly satisfactory and quickly declared secret, and landing speed was low, a fact which made it particularly suitable for landings on makeshift fields such as in the remote areas of China and New Guinea as would later be the case.

The first production B-25's had a constant wing dihedral. During the testing of the first B-25 at Wright Field, directional stability in the bombing run was not entirely satisfactory, and excessive dihedral in the wings was suspected as the reason for it. In correcting this the dihedral from the fuselage to the nacelle was retained, but from the nacelle out to the tip the dihedral was reduced to zero degrees. As a result, only the first nine B-25's had this dihedral, and all of the subsequent B-25's through the entire production run had the characteristic "gull wing."

The first B-25's were lightly armed, with only two rifle-caliber guns in the rear, the single gun in the nose, and the .50-cal. Browning in the tail. The maximum bomb load was 3600 lbs., and this was quite substantial. However, military thinking up to that time was based on the suposition that all a medium bomber needed for its own defense was a decent amount of burst speed. So, plans called for a lightly armed ship with a fair bomb load to drop in on a vital target, hit it, and run for home.

Just 19 days prior to the Army's contract for the B-25, war broke out in Europe on September 1, 1939. With the increasing role being played by air power in the Nazi's swift advanced across Europe, and with France and Britain now in the war, North American Aviation engineers placed emphasis on ease and rapidity of manufacturing being achieved to the highest possible degree.

A component breakdown plan was devised that allowed hundreds of suppliers and subcontractors to furnish equipment and subassemblies. By the end of B-25 production some 921 subcontractors had become involved. The company's original plan to divide the B-25 into 48 major component sections was never changed, except for some minor details, throughout the entire production run. Fisher Body Division of General Motors was the biggest provider of subassemblies.

It is fairly evident that once the wheels were up on this B-25A its clean lines and powerful engines would make it a very fast bombardment airplane.

B-25A

No more than 25 airplanes had been completed when the first change order came through calling for installation of the still experimental self-sealing gas tanks. This change resulted in the airplane becoming the B-25A.

The B-25A was the first operational model, and was delivered to the 17th Bombardment Group (Medium) in mid 1941, re-equipping from B-18's at McChord Field near Tacoma, Washington. Several months were spent on operational trials with the new airplane, and these first B-25's were well received.

As with the B-25, the B-25A carried three .30-cal. guns, and the single .50-cal. gun in the tail.

The eighth B-25A (AAC 42-196), and the 13th (AAC 42-201), became RB-25A's which classified

North American B-25A, representative of the early B-25 production.

them as "Restricted" and not to be used on combat missions.

The Navy displayed a geniune interest in the B-25A, but the Army Air Corps resisted these Navy efforts to secure large numbers of these land-based bombers, as they also did later with the B-24.

The B-25 entered World War II with little fanfare. At the time of Pearl Harbor the nearest B-25's

to the Pacific combat theaters were stationed along the Pacific Coast.

After Pearl Harbor, the 34th, 37th, and 95th Squadrons, which comprised the 17th Bombardment Group (Medium), were diverted to anti-submarine patrols along the Pacific Coast where their fire power was less important than bomb load and range. The airplanes were fitted with external bomb racks and increased fuel tankage to extend the time on station.

Sleek and slim, this early B-25A was completely different from any previous American bomber. The early open tail skid is visible.

B-25B

Army Air Corps' observers with the Royal Air Force kept reporting back on operations and tactics, and it was evident that changes would be required frequently on production lines to keep our planes abreast of and ahead of Axis combat types. When the relative speed of enemy fighters proved high enough, more armament would be needed on our bombers.

Power-driven turrets had proved their worth in the Battle of Britain in August and September of 1940. Electrically operated upper and lower turrets, each mounting twin .50-cal. guns and developed by Bendix, were now substituted for the side and upper waist guns on the B-25 and B-25A. The upper turret was a direct fire unit with the Plexiglas dome, while the ventral turret was sighted with a periscope and fired by a gunner kneeling inside the fuselage. This lower turret was fully retractable, with the guns fitting into slots in the fuselage.

The tail gun was removed since both turrets could effectively fire to the rear. A transparent cone was mounted in place of the tail gun for a prone observation position to the rear. A separate photography station was now located in the aft fuselage.

By now the War Department standardized armament in favor of the .50-cal. machine guns rather than the lighter .30 caliber in use on most military aircraft through the pre-war era. The only .30-cal. gun retained on the B-25 was in the nose, and the remaining 119 airplanes of the original contract were produced with this gun arrangement, and were designated as the B-25B.

In March, 1942, the first B-25's were being ferried over the ocean when a number were delivered to Australia.

It was the B-25B (North American's model NA-62B) that was used in the famous raid on Tokyo. The first six months of American involvement in World War II saw our forces systematically overwhelmed and our bases increasingly abandoned. The American people needed a morale boost desperately.

Along about this time, North American Aviation had a "Doolittle Project" which called for special installation to be made in the utmost secrecy. No one knew what this "Doolittle Project" was about until a few months later.

The B-25B was chosen over the Martin B-26 for this project because of its superior take-off performance. Volunteer crews were selected from the 17th Bombardment Group squadrons and underwent special training.

These B-25B's were modified for this operation by removing the lower gun turret, saving some 600 lbs. in weight, and installing 50-gal. fuel tanks in its place, which brought the total internal fuel capacity to 1141 gals., this to be added from five-gallon cans carried in the airplane. Normally, the B-25B would carry 694 gals. With this extra fuel load, the maximum gross weight was raised to 31,000 lbs. Special low-level bomb sights were built for operating reasons as the bombing altitude was to be only 1500 ft., and for security reasons so that the secret Norden bombsight would not fall into enemy hands. To discourage Japanese fighters from making tail attacks, two dummy wooden .50-cal. guns were

placed in the extreme rear fuselage. Most of this modification work was accomplished in February, 1942 by Mid-Continent Airlines in their Minneapolis shops.

Even with these high weights, under training by naval carrier experts, the crews were able to take off in 700-750 ft. during training.

Sixteen B-25B's were hoisted aboard the U.S.S. Hornet at Alameda Naval Air Station, the first twin-engined aircraft to be based aboard a carrier, and sailed for a secret destination. When 800 miles out from their Tokyo target, the bombers took off from the pitching deck with the lead plane flown by Lt. Col. James Doolittle. This launch was sooner than planned because of an encounter with an enemy patrol boat which could have alerted enemy forces.

Although the B-25's hit most of their targets, all of the airplanes were lost through bad weather and fuel shortages (caused by the earlier take-off) when attempting to find their designated bases in China. Most crews were saved, and the damage to Japanese targets not great, but it gave American morale a tremendous boost and also demonstrated the Japanese homeland's vulnerability to air attack.

Speculation had it that Doolittle had taken off from the mythical "Shangri-La." That they could have taken off from a carrier was considered by some as the only logical possibility, but the theory was readily rejected as unlikely because of the B-25's wing span and take-off needs.

Doolittle was awarded the Congressional Medal of Honor for his organization and leadership of the raid, and promoted to Brigadier General. Seventy-nine other volunteers who participated in the action received the Distinguished Service Cross.

One of the Tokyo bombers landed at Vladivostok and was confiscated by the Russians and its crew interned, this because Russia was not at war with Japan. The Russians were not strangers to the B-25 as some of their pilots were sent to the United States in late 1941 for training for the many B-25's that would be supplied under Lend-Lease.

Early in April, 1942, ten B-25's equipped with auxiliary fuel tanks accompanied three B-17's from Darwin, in Australia, to the Philippines where for several days they operated out of clandestine airstrips to raid Japanese-held docks, air and harbor facilities and other military targets, and suffered no combat losses. This first 4000-mile round trip outside blow to be made against the Philippines was organized and led by Brigadier General Ralph Royce, and was carried off against all the advice of experienced combat pilots.

At Nichols Field, north of Manila, they destroyed hangars and runways. At Batangas, a South Luzon port, they sank ships laden with military supplies. At Cebu three Japanese transports were sunk. Some 650 miles below Manila at Davao in southeastern Mindanao they destroyed a Japanese bomber, damaged several others, shot down two seaplanes, bombed two transport ships, and hit troop concentrations, docks and warehouses. These devastating sweeps caused eight Japanese ships to be sunk or damaged, and five aircraft to be destroyed. Evacuees from the beleaguered islands were also flown out. The B-25's of the 3rd Bombardment Group also began operations in early April against targets in New Guinea, and from that time onward the B-25 played an increasing role in air warfare.

The British Air Ministry received 23 B-25B's (RAF serials FK161 to FK183) as the **Mitchell I**, but these used only for familiarization and training, and never reached operational status.

A total of 1619 B-25C's had been built at the Inglewood, California plant through 1941, for a total of 1625 of this model. The B-25C was basically unchanged from the B-25B that raided Tokyo, including the same autopilots which were used on the raid, and the installation of Wright R-2600-13 engines.

Self-sealing oil tanks were also found to be as desirable as self-sealing fuel tanks and these were also included, and a larger escape hatch was provided to enable crew members to get out quickly. Other changes were found in radio equipment, and with superchargers and armor plate added the normal bomb load was now reduced to a mere 3000 lbs.

Shortly after the U.S. entry into World War II the B-25's were named **Mitchell** in a rather circuitous yet sincere effort to honor Brigadier General William "Billy" Mitchell, the great prophet of air power during the 1920's. The name originated with and was recommended by North American Aviation executives, rather than the military where

A standard B-25C-NA, AAC 41-12633, shows its dorsal turret and retracted ventral turret.

B-25C

Now the Army reordered the B-25 in an improved version, calling for 863 airplanes to be designated as the B-25C. North American, incorporating all the previous modifications and adding external bomb racks, now called this B-25C model their NA-82. In addition, the new contract called for an additional .50-cal. fixed gun in the nose, along with a flexible .50-cal. gun substituted for the .30-cal. weapon used by the bombardier. The fixed gun could be fired by the pilot who aimed the bomber at the target as would a fighter pilot.

many of the principals involved in the Mitchell affair were still on active status.

These airplanes were manufactured in several variations. Model NA-82 was the basic B-25C, of which 863 were built. The B-25C-5-NA was the company's NA-90 model and 162 were built for the Netherlands East Indies Air Force. The NA-93 model had its counterpart in 150 B-25C-15-NA bombers built as Defense Aid for China. The NA-94 became the B-25C-10-NA, only 150 of them being built as Defense Aid for Britain, and the NA-96 was built as 300 airplanes in the B-25C-20-NA and B-25C-25-NA versions.

As the **Mitchell II**, the Royal Air Force received 433 B-25C's. These were initially used in October, 1942 to equip Nos. 98 and 180 Squadrons in the No. 2 Group then in Bomber Command. Their first action took place January 22, 1943 with a raid on oil refineries in Belgium.

These units were joined by Squadrons 226, 305, 320, and 324 of the 2nd Tactical Air Force. No. 320 Squadron was manned entirely by Dutch personnel and became operational in July 1943.

The Royal Canadian Air Force also acquired four **Mitchell II's** on a cash-reimbursement basis.

B-25D

By now, North American was preparing to build the NA-73 **Mustang** for the British Government

The airplane covered by the manual accompanying this story is that of the B-25D model. The B-25D was the mid-point from which came all the many modifications and derivatives which made the **Mitchell** perhaps the most formidable aircraft of World War II.

The Navy and Marine Corps operated the PBJ-1D, basically a B-25D, but differing in having a tail turret with a single .50-cal. gun and waist positions as on the later B-25H. These PBJ's usually carried depth charges and bombs. Some of these hybrids were also supplied to the Royal Air Force which placed them in the **Mitchell III** classification. Some PBJ's had nose radomes, while others had wingtip radar.

The first B-25's to arrive in Russia to take part in the pre-Stalingrad actions of 1942 were B-25D's and

A well-worn Royal Air Force Mitchell II blends with the landscape below. This was the B-25C as supplied through Lend-Lease.

A trio of Netherlands East Indies' B-25C-5-NA's heads out looking for the enemy. The near ship has RAAF markings obliterated and replaced with wartime NEI markings.

and was still endlessly turning out AT-6's and **Harvards.**

The Inglewood plant could no longer keep up with the production demands and North American Aviation set up another plant at Kansas City, Missouri and built the B-25D, a total of 2290 which were built. The B-25D, for the most part, was basically identical to the B-25C except for the plant location and the short exhaust stacks which gave the B-25 its noisy reputation. Yet, 1200 of them built as the B-25D-1, -5, -10, and -15 blocks were North American's model NA-87, and the other 1090 were the model NA-100, built as the B-25D-20, -25, -30, and -35 blocks. All of these airplanes had "NC" after the military designation, denoting the Kansas City plant.

A former B-25C in service with the Royal Canadian Air Force as a Mitchell II, carries RAF serial KL149 and named "Li'l Abner," is a strange combination. It features two nose guns, bay windows, half-ring exhaust system and a plastic astrodome.

The Air Force Museum in Fairborn, Ohio has this B-25D-30-NC, AAF 43-3374, on permanent display. However, the airplane is painted to resemble a B-25B with serial AAC 40-2344. Seen here in two different paint schemes, the one above is its current status.

were immediately well received and liked, and went into action on July 27, 1942. The Royal Canadian Air Force also flew B-25D's as target tugs.

The basic versions of the B-25C and B-25D had a normal combat weight of 29,500 lbs., with a rarely used permissible take-off weight of 41,800 lbs. Along with the four main wing tanks for 670 gals., these models also had a bomb-bay tank holding 585 gals. for a total of 1255 gals. An additional 304 gals. could be carried in six auxiliary wing tanks, plus another 125 gals. in a waist tank, for a total fuel capacity of 1684 gals. for ferry purposes and a combat weight of 32,500 lbs.

In some of the early B-25C and B-25D variants, the lower turret was not fitted, but most production batches had provision for increased bomb stowage.

Alternate bomb loads for both models were one 2000-lb., two 1000-lb., six 500-lb., eight 250-lb., or twelve 100-lb. bombs in the bomb bay, except when additional tanks were installed for ferring purposes. The maximum offensive load for short-range operations was raised to 5200 lbs. by the addition of eight 250-lb. bombs on external racks. External racks appeared on the B-25C-1-NA and B-25D-1-NC series which permitted the carrying of a single short 22.4-in. torpedo weighing 2000 lbs. beneath the fuselage.

The B-25D had individual exhaust stacks attached to each cylinder with special exit fairings incorporated into the cowl skirt sections. Reports from military observers showed a necessity for dampening the exhaust flames as the airplane would be less visible at night.

Some of the late B-25C's and most of the B-25D's were rigged for a considerably shorter range so that they could carry a fixed and a flexible .50-cal. gun in the nose as well as the upper and lower turrets, and two and one-half tons of bombs as well, a combination which inspired the B-25G.

XB-25E, XB-25F

The XB-25E was the first and only one of two B-25's to carry an "XB" designation.

An experimental model of the B-25C, an NA-94, it was converted in 1942 from B-25C-10-NA, AAF 42-32281, and was fitted with a heated wing for hot-air de-icing purposes. It was subsequently operated as the SB-25E.

The XB-25F was converted in 1942, also from a B-25C model, for experimentation with thermal de-icers for the wings and tails.

B-25G

The B-25G was the first to carry a 75-mm. gun, the most spectacular model yet of the B-25 series.

The 90th Squadron of the 3rd Bombardment Group in the Pacific fitted some 75-mm. guns to their B-25's in the field. Experiments with a 75-mm. gun in airplanes were nothing new, but the major problem had always been an adequate system of range finding. Earlier, The Army Air Corps had proved the 75-mm. installation feasible in experiments in a Douglas B-18A. The cannon, with a special recoil device, was developed by Army Ordnance engineers.

Flown to combat areas with the cannon installed, the B-25's proved a surprise to the enemy which was not expecting to encounter "flying artillery." It was

The Timken Roller Bearing Co. of Canton, Ohio for many years flew this converted B-25D-NC as N-122B. Formerly AAF 41-29784, it was fitted with Patushkin wing-tip tanks, converted to a full collector-ring exhaust system. Extremely plush, it also incorporated an electric wheel-chair elevator in the aft fuselage.

With its nose guns and package guns removed, a B-25D settles in for a landing. The location of the package guns is still discernible on the bomber's side.

A PBJ-1D of Marine Squadron VMB-423 somewhere in the Pacific combat area is shown with the four package guns and two fixed guns in the nose.

In service with the Royal Air Force as the Mitchell III, this is a hybrid version of the B-25D supplied to the Marine Corps as a RBJ-1D.

something no other airplane in the world had been able to accomplish successfully.

The 75-mm. M-4 gun was actually of field-gun-size bore and weight, an airborne version of the French 75 which was proven so devastatingly effective by the Allies in World War I.

The mount was specially designed and restricted. The cannon was 9-ft. 6-in. long and weight was about 900 lbs., and it was mounted in a cradle beneath the pilot's seat in what was normally the tunnel to the nose. A hydro-spring mechanism formed part of the gun mount to take up the 21-in. recoil each time the approximately 3-in., 15-lb. shell was fired. The shell itself with its charge was 26-in. long and weighed 20 lbs. Each B-25 carried 21 shells which were loaded and fired by the navigator/cannoneer. The recoil was negligible, and pilots described the sensation of the 75-mm. cannon firing as similar to the shake of an auto coughing on a cold morning. The twin .50-cal. Brownings in the nose were fired simultaneously with the cannon during the firing run to suppress any enemy fire and for sighting on the target.

Test firing the cannon into a sandbank, engineers carefully calculated the stresses and allowed for them before the cannon-carrying plane was sent up for firing tests over the ocean. Little difficulty was encountered.

And again, the cannon installation was incorporated on the production lines at Inglewood. The B-25D nose was shortened and blocked in, with the cannon set in the lower left-hand side of the nose, and thus it became the B-25G-NA, 405 of which were ordered.

In addition to the 75-mm. gun, the B-25G also carried two fixed .50-cal. guns in the nose above the cannon, plus twin .50-cal. guns in the dorsal turret, all of these guns having 400 rounds each. The lower turret was not always fitted on the B-25G because hydraulic fluid and dust often obscured its sighting system, effectively rendering it so much dead weight. Cannon installations were also made at North American's large modification center in Kansas City.

Lieutenant General George C. Kenney, commanding the Far East Air Force, agreed to accept 63 B-25G's, and when they first arrived in July of 1943 they were tested chiefly in long-range operations against light naval craft and armed vessels and other Japanese targets by Lieutenant Colonel Paul I. Gunn. He found the cannon to be accurate enough during these operations, but it was impossible to fire off more than about four shells in each attack during which the airplane was extremely vulnerable to ground fire as no evasive action could be taken. More supporting fire power was deemed important to permit the aircraft to blanket the boats with .50-cal. fire during the runs to the target. To increase this offensive fire power, Gunn recommended the "wiring" of four .50-cal. fixed guns in individual external "packages" below the cockpit area.

Appearing stealthy and sinister, these two views of B-25G-10-NA, AAF 42-65128, show well the nose that carries the 75-mm. cannon and two fixed guns. The windshield has been changed to accept armor plating in the quarter panels for pilot protection, and the turret is the same as the B-25D.

After field modifications to provide these weapons, the blast during the firing caused considerable skin fractures in adjacent areas, and these areas were strengthened with flash plates at the Townsville Field Modification Center in Australia. Heavy aluminum sheets added alongside the package gun muzzles solved the skin scorching problems. Heavy aluminum plates also had to be added around the cannon muzzle because of scorch. These modifications were incorporated at the factory on subsequent aircraft.

Crude attempts were made to restore better rearward fire power such as "scare" guns in the rear of the engine nacelle, and swivel guns in the tail cone such as the old B-25A models had.

Some sources said that because of the recoil and resultant shock of the cannon to the cannoneer, the cannon was not too successful, and so it was used primarily against gun emplacements, tanks and other ground targets, along with shipping targets.

The B-25G was also sent to Italy in late 1943 for use in the Mediterranean area. It was useful for strategic shelling of such targets as power stations where putting the objectives out of action was more preferable than destroying them.

The B-25G arrived in Russia in time for the great counteroffensive of 1943 which routed the German armies.

The Royal Air Force had two B-25G's (Serials FR208 and FR209) for evaluation but these were never sent into action. The U.S. Navy also had a few B-25G's designated as PBJ-1G's.

Up to September, 1943, 175 B-25C and B-25D models had been converted for ground strafing with four guns in the nose. Although the B-25G used the big cannon against anti-aircraft positions, shipping and other targets, it was particularly useful in softening up New Britain were 1253 of the 75-mm. shells were expended by B-25G's in 36 days. In the Marshall Islands campaign of early 1944, it became increasingly harder to locate worthwhile targets and, in most cases, the 75-mm. cannons were removed and in their places were installed two additional .50-cal. guns in the nose, as was the case with the aforementioned B-25C's and B-25D's.

B-25H

Again, all the field modifications made to the B-25G were incorporated in a 1943 contract for 1000 more cannon-carrying bombers, this model to be the B-25H—the North American NA-98 model. Variants of the airplane were B-25H-1-NA to B-25H-10-NA.

It differed from the B-25G in having a newer and lighter T13E1 type 75-mm. gun, and the remaining

With the cannon removed and the outlet covered, this B-25G was a wartime conversion to a command airplane for a Brigadier General.

In olive-drab paint, this B-25H-1-NA, AAF 43-4198, shows off the stubby nose of the cannon-carrying version of the B-25.

The B-25H-1-NA, AAF 43-4112, came off the lines usually unpainted although this one is seen in full paint. To make it different, the upper turret has been covered over, it has an unarmored windshield and the nose gear nutcracker is covered by a boot. The heavy flash plates for the package guns are easily visible.

armament was retained and augmented. Except for a few early B-25H-1-NA's, four .50 cal. guns in the nose with 400 rounds each, were mounted along with the now standardized four .50-cal. package guns, two on each side of the armored nose. The dorsal turret was moved to a more forward position just aft of the cockpit. The ventral turret was deleted in favor of a new arrangement of rearward-firing armament. Sockets in curved bay windows in the waist now provided mountings for a .50-cal. flexible gun on each side, with 200 rounds per gun, for downward and rearward coverage. A position for a tail gunner was again included with twin .50's in a power-operated mounting, each supplied with 600 rounds and fired from a comfortable sitting position. The 15 guns in the B-25H constituted the greatest number of guns ever packed into a medium bomber.

When the four nose guns and the cannon were fired at once, tremendous pressure would build up inside the nose and the cover would fly open, necessitating a field modification adding a small vent on the nose. This, too, was included on the production line. The eight forward-firing guns, augmented by the two in the top turret, all firing in support of the 75-mm., was just awesome, and assured the beating down of any defensive deck fire on anything encountered all the way up to light cruisers.

To differentiate from the B-25G model, this **B-25H-5-NA, AAF 43-4570,** shows how the dorsal turret has been moved forward, the addition of two more guns in the nose as well as the package guns, the side bay windows, and the tail gun emplacement.

The new B-25H's were coming off the line unpainted, and these would be camouflaged in the field as required.

The B-25H could also carry up to 3200 lbs. of bombs or a 2000-lb. torpedo. Its five-man crew consisted of pilot, navigator-radio operator-cannoneer, flight engineer-dorsal gunner, mid-ships gunner-camera operator, and the tail gunner.

The first B-25H deliveries to the Far East Air Forces were in February, 1944, entering service with the 498th Squadron. However, the heavy armament was conferring the same disadvantages as the B-25G, and was abandoned in the Pacific combat areas by August, 1944. Some of the B-25H's replaced the 75-mm. gun with two rockets mounted piggy-back in the cannon location with a third rocket slung under the left side of the nose. By that time, the 509 B-25's in the Far East Air Forces outnumbered all other types.

As the PBJ-1H, the Navy and Marines flew the B-25H. It also proved successful in Russian service and repeat orders were placed for the airplane. Some of the Russian versions were modified to take six guns in the nose along with the 75-mm. cannon.

In natural finish, this **B-25J-20-NC, AAF 44-29259** displays its fire power. However, the package guns are not mounted on this particular airplane.

B-25J

In the B-25J, production reverted to the bomber nose of the B-25D, a strong reminder that the **Mitchell** was still basically a bombardment airplane. Even so, it carried two fixed and one flexible gun in the nose, the four guns in the fuselage package, the two waist guns, and the twin guns in the upper turret and the tail position, a total of 13 .50-cal. Colt-Browning machine guns.

The B-25J was the most widely produced of the line, with variants from the B-25J-1-NC to B-25J-35-NC. This was North American's model

Bristling with armament, this **B-25H-5-NA, AAF 43-4550,** shows off its fire power. Note the **armor plating on the windshield.**

Viewed head on, this **B-25J** displays the distinctive gull-wings that characterized the Mitchell. The package guns can be seen above the open bomb bay.

NA-108. A total of 4318 were turned out at Kansas City between 1943 and 1945.

A six-man crew was carried by the B-25J, but otherwise the layout was the same as the B-25H.

The maximum bomb load was six 500-lb. bombs or three 1000-lb. bombs. The 2000-lb. bomb station was deleted after 150 B-25J-1-NC's were built. Maximum speed of the B-25J at normal combat weight in 1945 was 285 m.p.h.

Upon arrival in the Pacific in the Spring of 1944,

the B-25J soon proved unsuited to conditions in that combat theater, where medium bombers operated almost exclusively at low altitudes. The B-25 had by then generally replaced the Martin B-26 as the standard medium bomber in the Pacific.

Low-altitude techniques did not require the services of a bombardier, and some of the B-25J's were modified to an eight-gun nose configuration. The bombardier noses of the B-25J's were replaced with a solid molded nose kit supplied from the factory. But, from September, 1944 on, the B-25J's with "attack" armament instead of the bomber nose arrived in the Pacific direct from the factory. These also

A typical B-25J-1-NC, this one AAF 43-3889, with a full complement of guns, yet in full paint. Also shown is another B-25J-1-NC, AAF 43-3892, in identical configuration and a few spots behind the other on the production line, yet left in its natural metal finish.

were altered in the field by the addition of a 150-gal. fuel tank in the radio compartment. Still later B-25J's were the most heavily armed of all with 18 .50-cal. guns, and were used principally in the Pacific where underwing stubs were also fitted for carrying four 5-in. rockets outboard of each nacelle.

As the **Mitchell III**, 314 B-25J's were delivered to the Royal Air Force, and these were used principally with the 2nd Tactical Air Force. These were in RAF serial ranges HD346-HD400, KJ561-KJ800, and KP308-KP328. The Royal Canadian Air Force used 100 B-25J's with the bomber nose for all-weather crew training and guided-weapon development.

A B-25J undergoes under the most primitive of conditions at a forward base in the Pacific combat area. Unlike other J models, this one has only one fixed and one flexible gun in the nose.

The Marines employed the PBJ-1J, which had minor differences in radio and other equipment from the B-25J.

The B-25J was supplied to many countries in 1947 and 1948 under the Mutual Defense treaties. These airplanes served as first-line attack-bomber equipment with some countries for many years.

Brazil had two light-bomber groups, outfitted with 29 B-25J's supplied under Lend-Lease, and was still flying them on operational status as late as the early 1960's.

Canada flew the TB-25J, as well as the TB-25D well into the 1960's, and Cuba used the B-25J in the pre-revolutionary era and probably after as well.

The Nationalist Chinese Air Force operated B-25's in the postwar era, as did Mexico.

Uruguay received 12 B-25J's in 1960, and Venezuela equipped one squadron with that type.

The B-25J's were also used in Bolivia, Chile, Indonesia, and Peru. With the many military coup's that took place in these countries, it is highly likely that the B-25J's weren't used for peaceful purposes.

A Mitchell III of the Royal Canadian Air Force plies its daily courier and mail "milk run" above the Arctic Circle. The turret and package guns have been removed and crudely patched over. External wing racks can be seen, and it still has the armored windshield.

Once a B-25J-32-NC, this solid-nose model was converted to a TB-25N-30-NC, as were glass-nose versions, and is AAF 44-31334A. It has the half-ring exhaust system, clear windshield, but also has larger box-type carburetor intakes.

Once a B-25J-1-NC, AAF 43-4030 was specially modified as a personal airplane for Gen. Dwight Eisenhower. As a CB-25J, it had the bomber nose covered with metal and special windows added to the rear fuselage. The airplane is shown here as reassigned to a two-star general.

TRANSPORT VERSIONS OF THE B-25

Between 1942 and 1949, six B-25's were modified by North American to RB-25 personnel transports. Two were converted for company use, three for generals, and another as a combination prototype for a military trainer/transport or an executive airplane.

In December, 1942, the first B-25 built, AAC 40-2165, was taken off flight status and brought to North American's experimental department for modification. After removal of all military equipment, passenger accomodations were installed, including five seats in the aft fuselage plus a desk and intercom system. The bomb bay was converted for baggage stowage and a bunk was installed over it. Completed in April, 1943, it was referred to as **Dutch's Express** or **Whiskey Express**, and served

This RB-25 photo-reconnaissance airplane, formerly an F-10, is seen here still in service in 1951, and shows the chin fairings for the trimetragon cameras. The airplane is undergoing a thorough scrubbing down.

as North American Aviation's corporate transport for 2-1/2 years before it crash landed at the factory and was scrapped.

When General Henry H. Arnold, Army Air Corps Chief, visited the plant in mid 1943, he took more than passing notice of the B-25 transport and initiated action to have one converted for his personal use. At this his request, a B-25C was diverted from final assembly to the Field Service Department and modified as was 40-2165. All the work was done with government funding, and it was reportedly later owned for nearly 20 years by the late Howard Hughes.

Then, on March 1, 1943, B-25J-1-NC, AAF 43-4030, was flown to the Inglewood facility to be converted to a personal airplane for General Dwight D. Eisenhower. Work was completed on May 11 under a standard sales order. Rather than the quilted olive-drab "skyfelt" material so common in military aircraft, it was upholstered in blue fabric. Its solid nose included clam-shell doors for easy access to radio and navigation equipment. The rear entry hatch was moved further aft to provide an increase in usable floor area, and luggage racks were installed above the seats. The de-icer boots were retained and a long-range ferry tank was fitted into the top of the bomb bay above the baggage area. It was first designated as an RB-25J, then later the CB-25J.

In 1944, General Arnold turned in his RB-25C for a newly converted RB-25J-15-NC, AAF 44-28945,

Typical of the pilot-training conversions is this former B-25J-15-NC, now a TB-25N-15-NC, AAF 44-29043. It still retains the short exhaust stacks, but has a clear windshield.

almost identical to Eisenhower's and also modified by the Field Service Department. Arnold used the airplane until January, 1946 during which time it was alternately designated B-25J and RB-15J. It survived to see Air Force duty into the 1950's as did Eisenhower's.

The crash of the company airplane left North American Aviation without a transport, and conversion of a B-25J-25-NC, AAF 44-30047, was begun in January, 1945. Another transport was an immediate company requirement.

But, as the end of the war neared, the configuration was considered appropriate for development of a new idea that had been in management's mind for some time. The interior arrangements and appointments were generally the same as the previous two conversions, but other revisions were made. A

Once the flagship of the American corporate aircraft fleet, this converted solid-nose B-25J-35-NC, N-75754, was AAF 45-8882 but never saw military duty as it was not delivered yet when the war ended. It went through a number of refinements, and is shown here with a double half-ring exhaust system.

Fairly typical of the Mitchell in its immediate post-war operational role is this B-25J-25-NC, AAF 44-31027, with all armament removed.

heater was installed in the nose under the radio gear, and a direction finder was located in the same compartment. Access to the radio equipment was simplified by using the standard eight-gun nose without the gun ports. Large doors on each side provided easy access for maintenance. The airplane was completed in mid March and the test program began for CAA certification with the post-war transport market foremost in mind. Between company flights, certification tests were flown. On February 27, 1946, while on a routine certification flight, North American test-pilot Joe Burton radioed a report of fire in one wing. The aircraft was seen to explode over the ocean and North American lost one of its best test pilots and crews.

The sixth transport project undertaken by North American Aviation was some five years later. A B-25J-25-NC, AAF 44-30975 (PBJ-1J, BuA 35848) was purchased from surplus and designated by the company as "Executive Transport." The entire nose

This VB-25J-30-NC, AAF 44-30971, unaccountably has the same aft fuselage modifications made to the B-25's used by Eisenhower and Arnold, but is otherwise a solid-nose B-25J-30-NC assigned to a three-star general.

As a TB-25J-35-NC, AAF 45-8846 is still fairly typical of a B-25J except for the large radome under the rear fuselage.

A solid-nose B-25J-32-NC, AAF 44-31321 was converted to a CB-25J-32-NC. It has the half-ring exhaust system and a clear windshield.

Modified as a radar fire-control trainer, this B-25J-25-NC became TB-25K-25-NC, AAF 44-30489, and was assigned to the North Dakota Air National Guard. The bombardier nose has been replaced by a radar nose and equipment.

forward of the propeller line was redesigned so that it was nearly 40 in. longer, and the cockpit was widened to 70 in. for comfort. Pilots' seats were moved forward and four more seats were added forward of the bomb bay. This arrangement provided excellent passenger accommodations for business, and as a military trainer it would permit an instructor to have three students in the cockpit instead of only one. The big nose hinged at the top like the B-25H nose. The bomb-bay doors were removed and the area was skinned over, the forward end fitted with a loading hatch and an electrically operated baggage elevator. A Convair 240 windshield assembly was fitted, and the Clayton exhaust stacks of the upper cylinders were removed and replaced by a half-collector ring to reduce the terrible B-25 noise level. The B-25 with the short stacks was one of the noisiest airplanes ever built, and the noise level with this modification was reduced to below that of a DC-3.

The project was completed in 1950 and it was dubbed the **Bulbous Nose B-25**, with the civil registration N-5126N. Test flights proved it to fly like any other B-25 despite the longer nose. A sales tour was organized to show the airplane to both commercial and military interests. On March 25, 1950, the airplane entered a storm front and extreme turbulence east of Phoenix and crashed near the town of Chandler, Arizona. Both wings failed in upward flexing, broke off and severed the tail, and the wreckage was scattered over a wide area. The loss of seven North American Aviation people was a

hard blow to the company, and the project was abandoned.

Before North American developed the ill-fated Executive Transport, many private conversions were carried out in the immediate post-war era, some of them most plush. Others were adapted for a variety of special applications, and still others became military command transports.

One of the best-known applications of the B-25 was that made by Paul Mantz. His B-25 was specially modified for a conglomeration of cameras for movie work, most notable of which were the special cameras mounted in the nose for filming with the **Cinerama** technique.

A meat-packing company in Bolivia used a fleet of B-25's for hauling meat out of the back country.

In the 1960's and 1970's many B-25's were adapted for fighting forest fires, a job at which they were and still are quite effective.

One of the best-known civil conversions of a B-25 is that of the late Paul Mantz which had a special nose to house the cameras used to film in the Cinerama technique, and many other motion-picture assignments. As N-1203, it was converted from a B-25H-5-NA, AAF 43-4643.

This TB-25N-25-NC, AAF 44-29952, was modified by Hayes for pilot training from a B-25J-16-NC, and has the upper-half exhaust-collector system.

Assigned to the USAF Instrument School, this TB-25N-15-NC, AAF 44-30715, has the half-ring exhaust system, and is a former B-25J-25-NC.

Converted for pilot training, this TB-25N-15-NC, AAF 44-29060, was formerly a B-25J-15-NC. Unlike other conversions, this one still has the short exhaust stacks.

A TB-25J-16-NC, AAF 44-28844 is a pretty straight-forward unarmed model built as a B-25J-15-NC.

Last, more than a few weary B-25's were used for smuggling illegal substances into the United States, and smuggling guns out.

TRAINER VERSIONS OF THE B-25

Some of the earlier B-25's were stripped of some operational equipment and used for air-crew training, with a total of 60 assorted B-25's converted to AT-24 advanced trainers.

The AT-24A-NC was converted in 1943 from a B-25D-NC; the AT-24B-NA, also converted in 1943, was a B-25G-NA; the AT-24C-NA was converted from a B-25C-NA in 1943; and, in 1944 the B-25J-NC was converted to the AT-24D-NC. When the Advanced Trainer (AT) designation was dropped in the late 1940's, the designations were changed to TB-25D, TB-25G, TB-25C, and TB-25J respectively.

By 1944 some 600-plus B-25J's were stripped of

Whereas the TB-25K was used to train crews to fly and operate the gunnery system of the F-89, the TB-25M was used for training in still later F-89 models with more sophisticated fire-control systems. This TB-25M-20-NC was formerly B-25J-25-NC, AAF 44-30444N, and was assigned to the Wisconsin Air National Guard. The bomber greenhouse has been mostly covered, and an astrodome is atop the fuselage. Large air intakes and half- ring exhaust systems are also part of the changes.

guns and other armament and became TB-25J's. One-hundred such conversions were delivered to the Royal Canadian Air Force.

By 1950 the Air Force still had approximately 1400 B-25J's in useable condition. North American proposed to modify these in four versions: pilot trainer, navigation trainer, radar-bombing trainer, and ten-place personnel transport, all at an estimated cost of $100,00 per plane. To this end North American built up the Executive Transport, as was mentioned earlier. The B-25 was a proven, dependable, easy-to-fly and stable bombing platform. The

Tactical Air Command Headquarters flew this TB-25J-25-NC, AAF 44-86800, in full paint. Formerly a B-25J-30-NC, it has the solid gun nose.

Witnessing the inconsistency of Air Force markings, this former B-25J-30-NC, AAF 44-30996, was converted simply as a TB-25. It has the half-ring exhaust system, and a squared-off tail cone.

alternative to such modification was the half-million-dollar Convair T-29 which was being purchased "off the shelf."

Even though North American Aviation lost interest in the modification project after the loss of the Executive Transport, obviously the Air Force had not, and a less comprehensive modification program was undertaken which gave the B-25J's a new and long lease on life. Hughes Tool Company of Tucson, Arizona and Hayes Aircraft Corp. of Birmingham, Alabama overhauled and modified B-25J's for various training roles and as Air Force command and executive transports.

As "Gabriel Dumont," this TB-25J, as RCAF 5267, still mounts the package guns and two nose guns, but an astrodome instead of the dorsal turret, and the half-ring exhaust.

A TB-25J as Royal Canadian Air Force 5237 operates from Saskatoon. A solid-nose version, it has the half-ring exhaust system.

As a TB-25K, AAF 44-30133N was converted from a B-25J-25-NC, and assigned to the Wisconsin Air National Guard for radar training. It carries nose radar, the half-ring exhaust, and clear windshield as well as the larger air intakes. The ship was use to train F-89 crews.

In 1951, 117 B-25J's were converted to TB-25K-NC radar trainers for the E-1 Fire Control systems.

Ninety more B-25J's were converted by Hayes in 1951 to TB-25L-NC's for pilot training. New seats and other items were installed as required by the new training directives.

In 1952, the newer E-5 Fire Control radar required the modification of 40 more B-25J's as the TB-25M-NC, with the work being done by Hughes.

Finally, another 47 were modified in 1954 by Hayes as the TB-25N-NC. These were like the TB-25L, but were fitted with Wright R-2600-29A engines.

PHOTO VERSIONS OF THE B-25

The F-10 photo airplane was the B-25D modified in 1943 to photo reconnaissance duties. The guns were removed and numerous special openings were provided for a wide variety of cameras, including a tri-metragon camera in the nose with chin fairings.

Ten such airplanes were converted in this fashion.

Extra fuel was also carried in the airplanes, which were used almost exclusively for photo-reconnaissance training. The R-2600-13 engines were replaced by R-2600-29's, and combat weight was restricted to a lower figure for increased speed.

The F-10 was later redesignated as the RB-25D.

B-25's IN ALLIED SERVICE

The Royal Air Force flew innumerable B-25 sorties over the English Channel, attacking German territory for the first time in September, 1944.

After the German invasion of the Soviet Union in June, 1941, Russia asked for Lend-Lease Aid as a claimant. President Roosevelt proclaimed Russian resistance to the German invaders as vital to the security of the United States and directed that the USSR be given top priority in allocation of machines. In August, 1941, the Russians were promised five B-25's.

"Pert'nint Poop" is the name bestowed on this F-10 photographic Mitchell. Eyes, complete with lashes and eyebrows grace the camera fairing.

Delivery of the first batch of 72 B-25's to Russia began in March, 1942. Pan American Airways crews flew them from Miami to Africa and then up to Habbaniya in Iraq, then to Teheran, Iran where they were picked up by Russian pilots and flown over the Caucasus Mountains to Red air-force units. By the end of 1942, 102 B-25's had been flight-delivered to the Russians, where they quickly became extremely popular. A total of 870 B-25's were supplied to Russia, of which 862 arrived at their destination. All but five of these were flown out, the other five — possibly the original five promised — were shipped to Northern Russian ports by sea. Only 128 were flown over the South Atlantic Ferry Route to Iran, and the remainder delivered over the Alaska-Siberia Ferry Route where the Russian pilots took them over at Fairbanks.

The B-25G arrived in Russia in time for the great counter-offensive of 1943. The **Mitchells** supplied to the USSR under Lend-Lease included B-25B, B-25C, B-25D, B-25G, and B-25J models. A number remained in service during the immediate post-war years and were given to NATO code name **Bank**.

A Netherlands Purchasing Commission placed a contract for the B-25 and it was the first export order for the design, and called for 1942 flyaway delivery. These were produced along side Army Air Corps, RAF, and Soviet B-25's. The Dutch Air Forces in the Netherlands East Indies had been promised some **Mitchells** in late 1941, but the situation in the South Pacific was so desperate after that time that the airplanes were taken over by the 13th and 19th squadrons of the 3rd Bombardment Group of the Army Air Corps operation in Australia.

In 1944, 32 B-25D's were accepted on behalf of the Dutch government, and were later transferred to the Royal Australian Air Force under their serials A47-1 to A47-25, and A47-33 to A47-39.

Eighteen B-25J's were also supplied to the RAAF, seven of these originally accepted on behalf of the Dutch government. These were A47-26 to A47-26 to A47-32, and A47-40 to A47-50.

The remnants of the Netherlands East Indies Air Force which reached Australia reformed into No. 18 Squadron of the RAAF. Eventually, the Netherlands East Indies received 249 B-25's under Lend-Lease.

Under Lend-Lease, China was sent 131 B-25's which were assigned to the 1st Bombardment Group (Medium) of the Chinese-American Composite Wing formed in Karachi, India in October, 1943.

NAVY AND MARINE CORPS B-25's

The U.S. Navy operated several hundred B-25's, with most of them assigned to the Marines. These retained their Army Air Corps' model numbers in association with the Navy designations. Thus, the PBJ-1C was the B-25C, the PBJ-1D was the B-25D, the PBJ-1G the B-25G, the PBJ-1H the B-25H, and the PBJ-1J the B-25J.

A navy PBJ-1, under the command of Lieutenant Louis M. Abernathy, was the first to send a U-boat to the bottom off the East Coast, dispatching it with depth charges.

FINE-TUNING THE DESIGN

Such a varied career for the B-25 would not have been possible had North American's engineers permitted the B-25 design to remain static. The engineering staff was insistent from the very beginning that the NA-62 be capable of maximum performance and serviceability, lend itself to rapidity of construction, and be a pilot's plane.

Over 500 change orders were put into effect, and by the start of 1943, the B-25's on the production lines were some 22,000 drawing changes better than the B-25's that raided Tokyo. However, throughout its production history the basic B-25 airframe itself was little changed from the first B-25.

The constant addition of fire power, armor, and equipment brought it from a lightly armed medium bomber flying under heavy fighter protection to where it could take care of itself without question, and take on a naval destroyer or armed merchantman as easily as a fighter plane. In 1941 the normal bomb loads were about 2400 lbs., and rose to 4000 lbs. in 1945.

In efforts to increase the range, large fuel tanks were designed and carried in the bomb bay, the bombardier's compartment, in the crawl tunnel beneath the cockpit, and in the aft sections of the fuselage, enabling the B-25's to be flown long distances directly to the fighting fronts, and to participate in long overwater missions. Later, outboard self-sealing tanks replaced all but the bomb-bay tanks, permitting the easy crossing of the Atlantic and Pacific in island-to-island hops.

When B-25's were sent to Alaska to bomb enemy positions on Kiska and Attu Islands, where the cold was extreme and the plane's regular heating system not sufficient, provision was made for electrically heated clothing. Alaskan operations also showed a need for winterization shutters which were designed and installed.

Long overwater hops became commonplace, and the installation of Sperry auto-pilots was ordered as standard equipment.

Defrosting tubes for the windshield, bombsight window, and navigator's window were installed.

Warm air was piped to the gun breeches to prevent the armament from gumming up in the extreme cold.

Electrical bomb-release mechanisms were substituted for the original mechanical releases.

B-25's were equipped with special devices that enabled them to carry a standard torpedo, and not long after reports of B-25's sinking Japanese shipping came back from the Pacific.

To conserve critical materials, wooden bomb-bay doors were designed, tested, and included on production aircraft. Wood seats were also installed for the bombardier and radio operator for the purpose of saving precious aluminum-alloy materials.

Sealed-beam landing lights were adapted for use in aircraft and were installed for better night-landing vision.

Carburetor de-icers were designed to combat ice-forming conditions.

Windshield wipers had to be installed on bombers operating in torrential rains in the Pacific combat areas.

Electric gun heaters were adapted to insure the proper functioning of the .50-cal. guns in freezing weather.

North American Aviation Service Representatives in Africa asked the company to design special dust excluders for B-25's operating in blinding sandstorms. Landing-gear struts also required special protection against sand.

Camera windows were installed on each side ahead of the tail.

Rocket-assist take-off bottles were fitted under the wing outboard of the nacelle. Attack planes were fitted with a fixed type, and heavier planes with droppable units.

A B-25C-NA, AAF 41-12823, in standard olive-drab paint and markings of 1943.

Fluorescent lighting was adapted for the pilot's instrument panel, as were blackout curtains for the navigator and radio operator so they could perform their duties at all times.

Armor plating was provided to protect the pilot. Later on, armor plating was provided for other crew members as well.

SUMMARY

The **Mitchell** made its mark on every flying front of World War II.

Out of a total of 9815 B-25's produced, the peak Army Air Forces inventory for the type was only 2656 in July, 1944. All the rest were supplied to the Allies or expended prior to that time.

The simplification of jobs was carried forward in the B-25 production to a greater degree than ever before in aviation manufacturing history, and it enabled the company to train rapidly tens of thousands of inexperienced men and women to build the planes.

Cost-wise, the price per unit for the B-25 declined dramatically. From 1939-1941, the unit cost was $180,031. In 1942 it was $153,396; in 1943 it was $151,894; in 1944 it reduced to $142,194; and finally dropped to $116,752 each in 1945.

In widespread use as a borate-bomber in fighting brush and forest fires, the B-25 would still turn in a good day's work. Here N-9866C, a former B-25J-15-NC, AAF 44-28833, is fitted with a special bulk tank protruding from the bomb bay into which the crew is pumping s borate slurry.

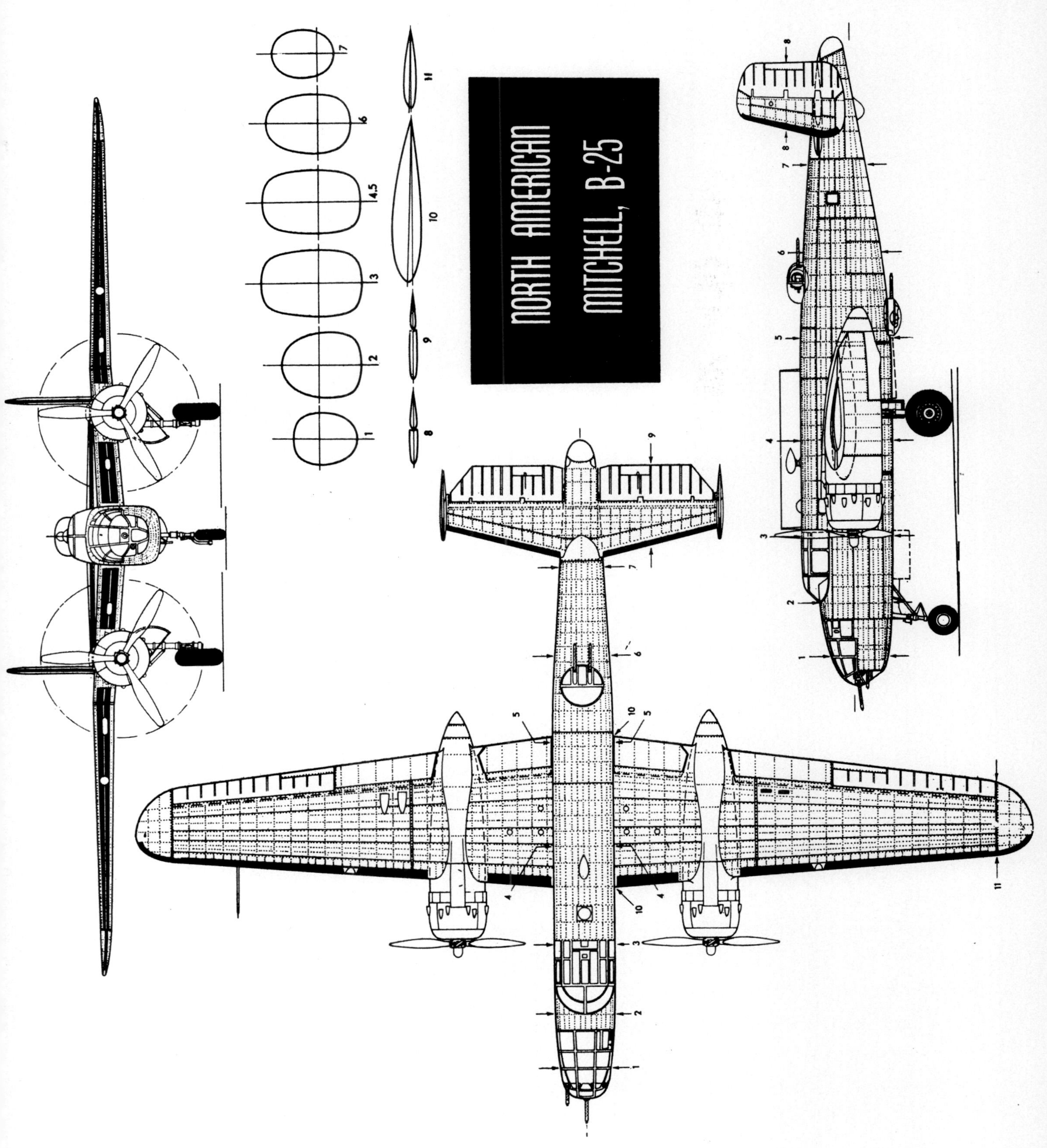

NORTH AMERICAN
MITCHELL, B-25

TABLE OF CONTENTS

ILLUSTRATIONS

Figure 1 - Exterior Views of B-25C and B-25D

SECTION I

DESCRIPTION

1. AIRPLANE

a. General. - The model B-25C and B-25D bombardment airplanes are mid-wing, land monoplanes, each powered by two Wright Cyclone R-2600-13 engines. Propellers are the three blade type, hydromatic, and full-feathering. Hydraulically operated tricycle landing gear, wing flaps, engine cowl flaps, bomb bay doors, and brakes are provided. The crew consists of the pilot, copilot, bombardier, radio operator, and photographer. The approximate over-all dimensions are:

Span	67 ft 6 in.
Length	53 ft
Height	15 ft 9 in

b. Landing Gear. - A fully retractable, hydraulically operated, tricycle type landing gear is provided. Doors cover the gear openings in both retracted and extended positions. The swivel type nose gear strut has a centering device which operates when the strut is fully extended (wheel off ground). An emergency air brake system is provided and consists of an air pressure storage tank in the navigator's compartment, an automatic transfer valve for each wheel, and an air valve controlled by a lever, accessible to pilot and copilot. The air pressure tank is provided with a gage and can be recharged from the ground only.

c. Indicators. - When either throttle is retarded for landing and the landing gear is not locked down, a warning horn will sound to warn of an unsafe landing condition. A release switch (figure 16-94) is provided to render the horn inoperative during one-engine or closed-throttle maneuvers. Opening the throttle automatically reinstates the warning horn. A position indicator (figure 16-111) for the landing gear, landing gear down-position locks, and wing flaps is provided on the pilot's instrument panel.

d. Hydraulic System. - Pressure for the system is supplied by two engine-driven pumps connected in parallel, one mounted on each engine. A hand pump for emergency operation is provided. (See figures 2, 3, and 4.)

e. Fuel System. - Each engine is provided with an independent system. The four main fuel tanks have a capacity of 658 U.S. gallons (548 Imperial gallons). A droppable bomb bay tank with a capacity of 418.8 U.S. gallons (348.9 Imperial gallons) is provided and is usable only as a reserve tank. Its fuel can be transferred only into the wing fuel compartments. (See figures 5 and 6.)

f. Fuel Transfer System. - (Applicable to B-25C airplanes, serial Nos. AC-41-12817 and higher, only, and B-25D airplanes, serial Nos. AC-41-29848 and higher only.) A fuel transfer system with electrically operated auxiliary transfer pumps and a manually controlled cross-feed shut-off valve is provided. (See figures 5 and 6.) No provision is made for returning fuel to the bomb bay tank.

g. Fuel Transfer System. - (Applicable to B-25C airplanes, serial Nos. AC-41-12434-12816 inclusive, only, and B-25D airplanes, serial Nos. AC-41-29648-29847 inclusive.) These earlier airplane models have a fuel transfer system with manually controlled selector valves and an electrically operated reversible pump. Fuel may be returned from the front wings to the bomb bay droppable tank. (See figures 5 and 6.)

h. Electrical System. - The electrical system is a 24-volt, single wire circuit. Two 24-volt batteries are provided, one inside each engine nacelle, immediately aft of the fire wall. Either battery operates the system, including starters. The batteries are charged by engine-driven 30-volt, 200-ampere generators. A generator control panel located at the right rear of the navigator's compartment controls the charging voltage of the generators.

i. Interphone System. - The RC-36 interphone system installed in these airplanes consists of a type BC-347 amplifier, one BC-366 jack box for each station, and one T-30 throat microphone for each crew member. The power for this equipment is furnished by a PE-86 dynamotor. For detailed radio operating instructions, refer to section V.

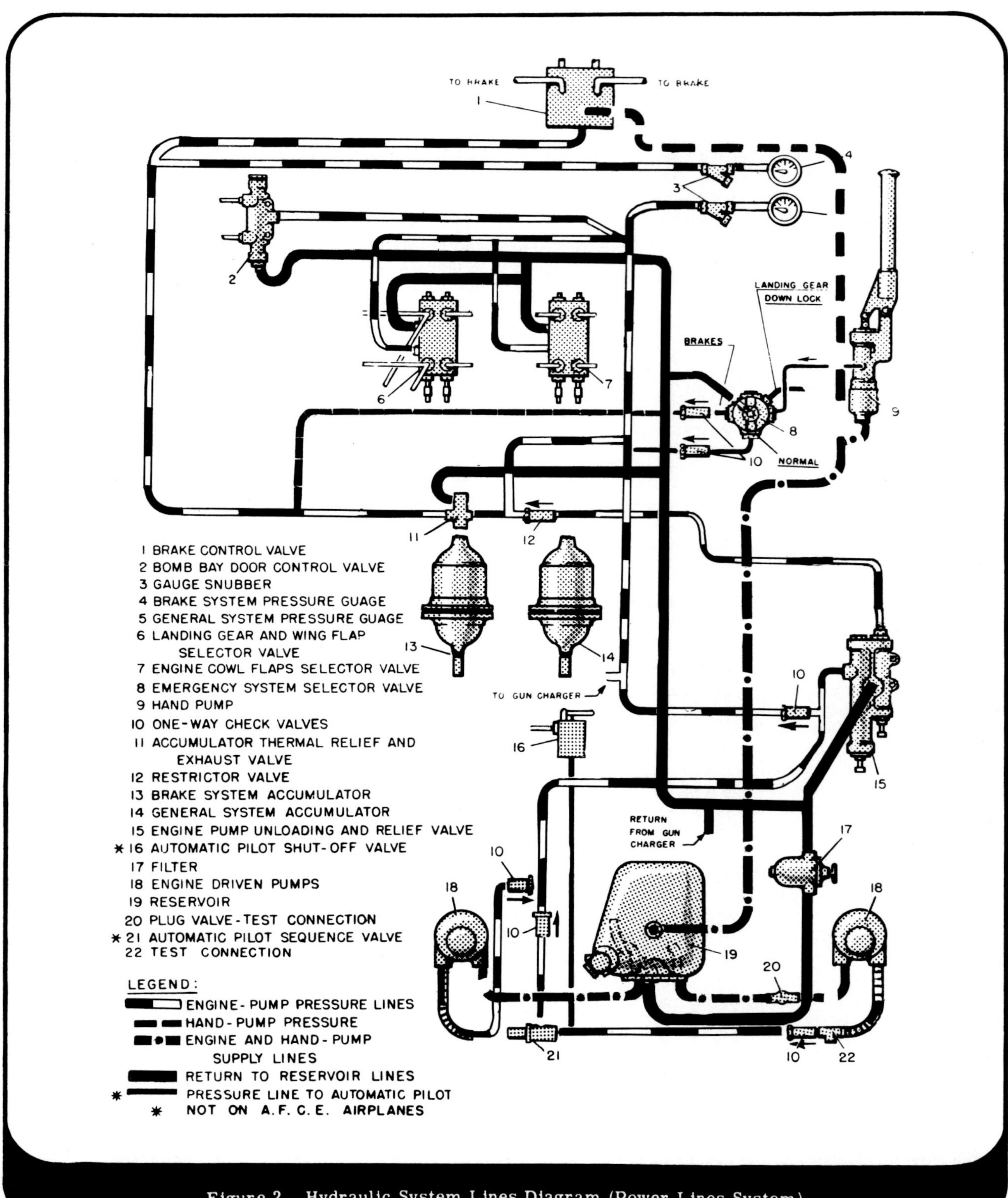

Figure 2 - Hydraulic System Lines Diagram (Power Lines System)

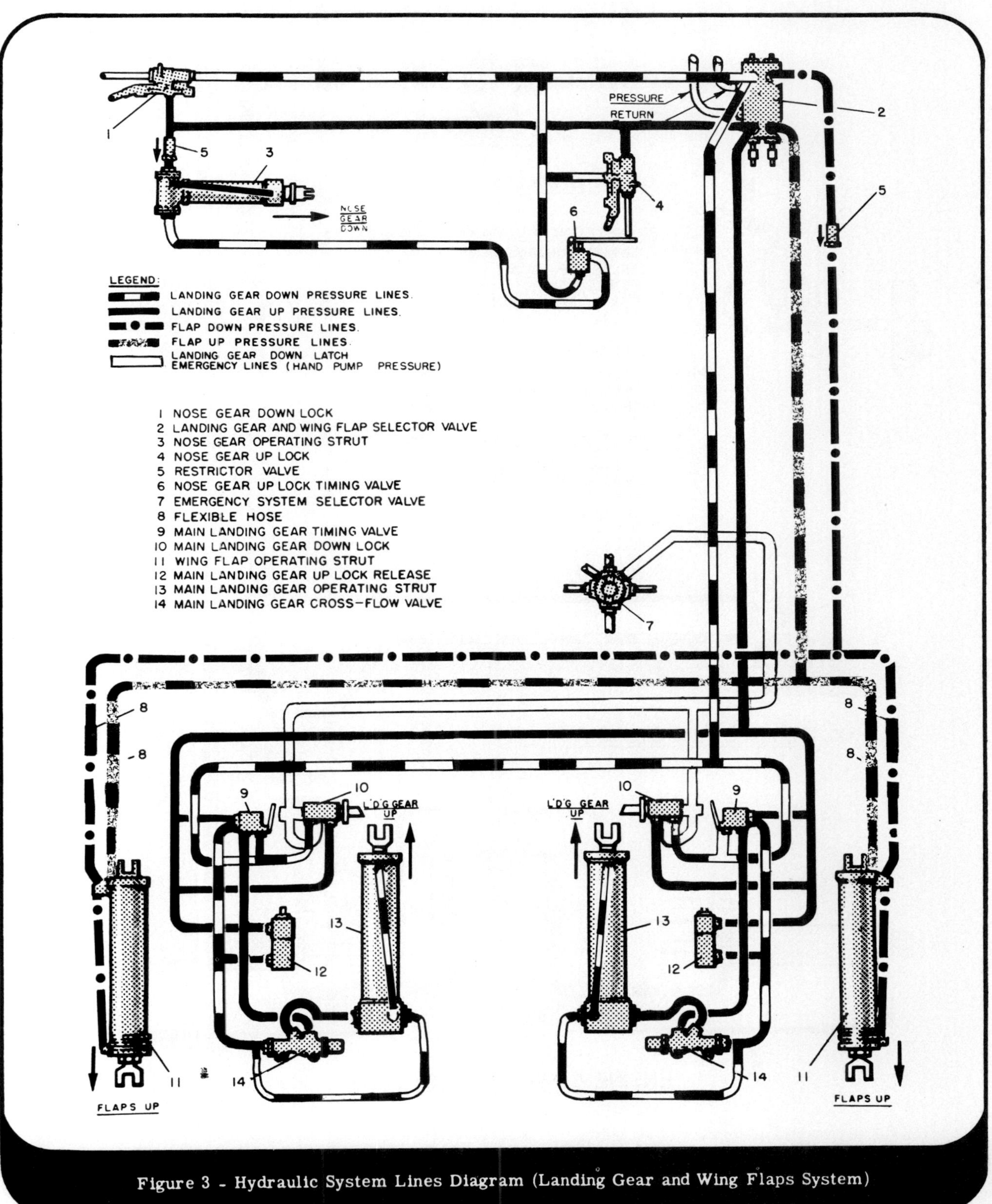

Figure 3 - Hydraulic System Lines Diagram (Landing Gear and Wing Flaps System)

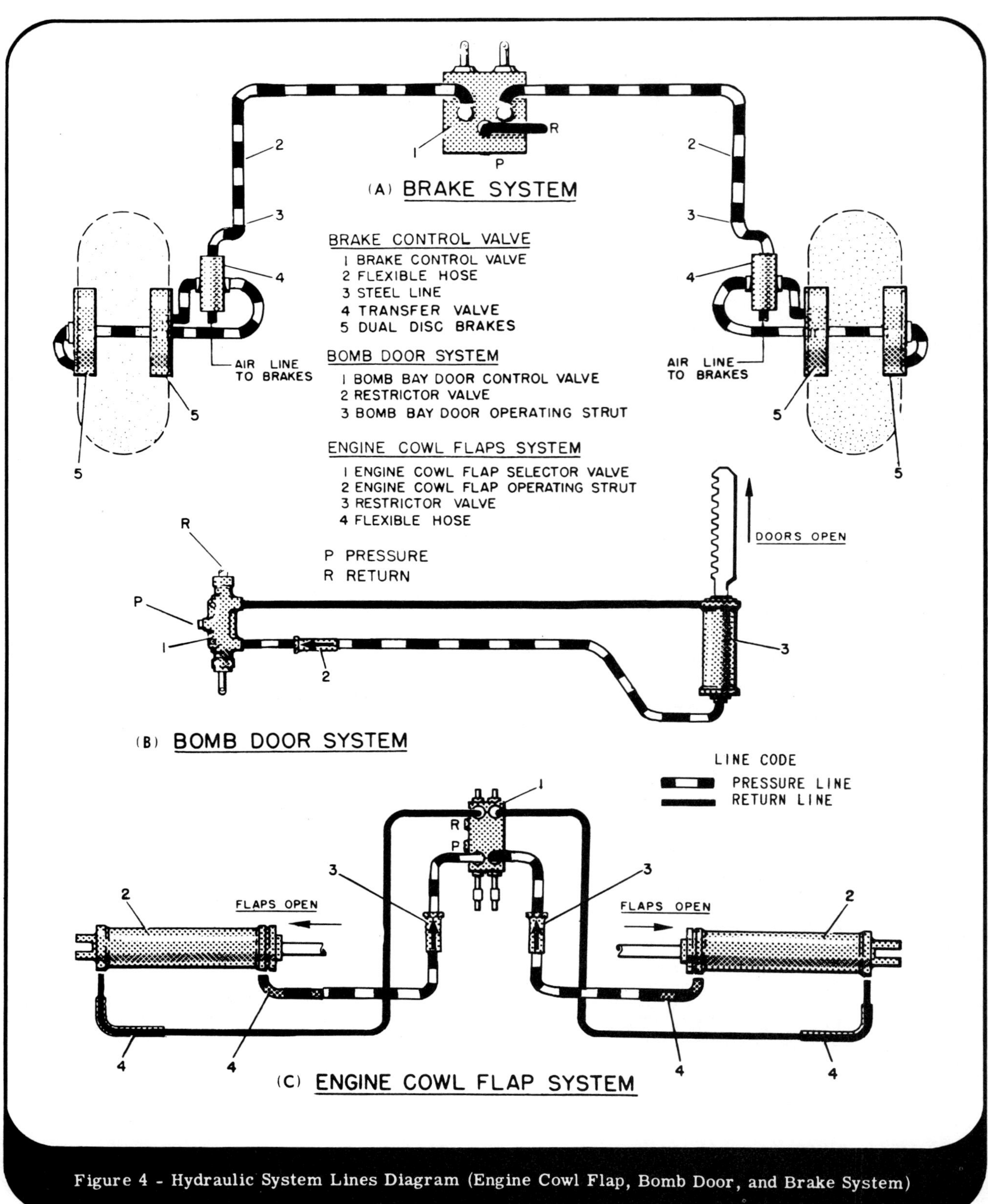

Figure 4 - Hydraulic System Lines Diagram (Engine Cowl Flap, Bomb Door, and Brake System)

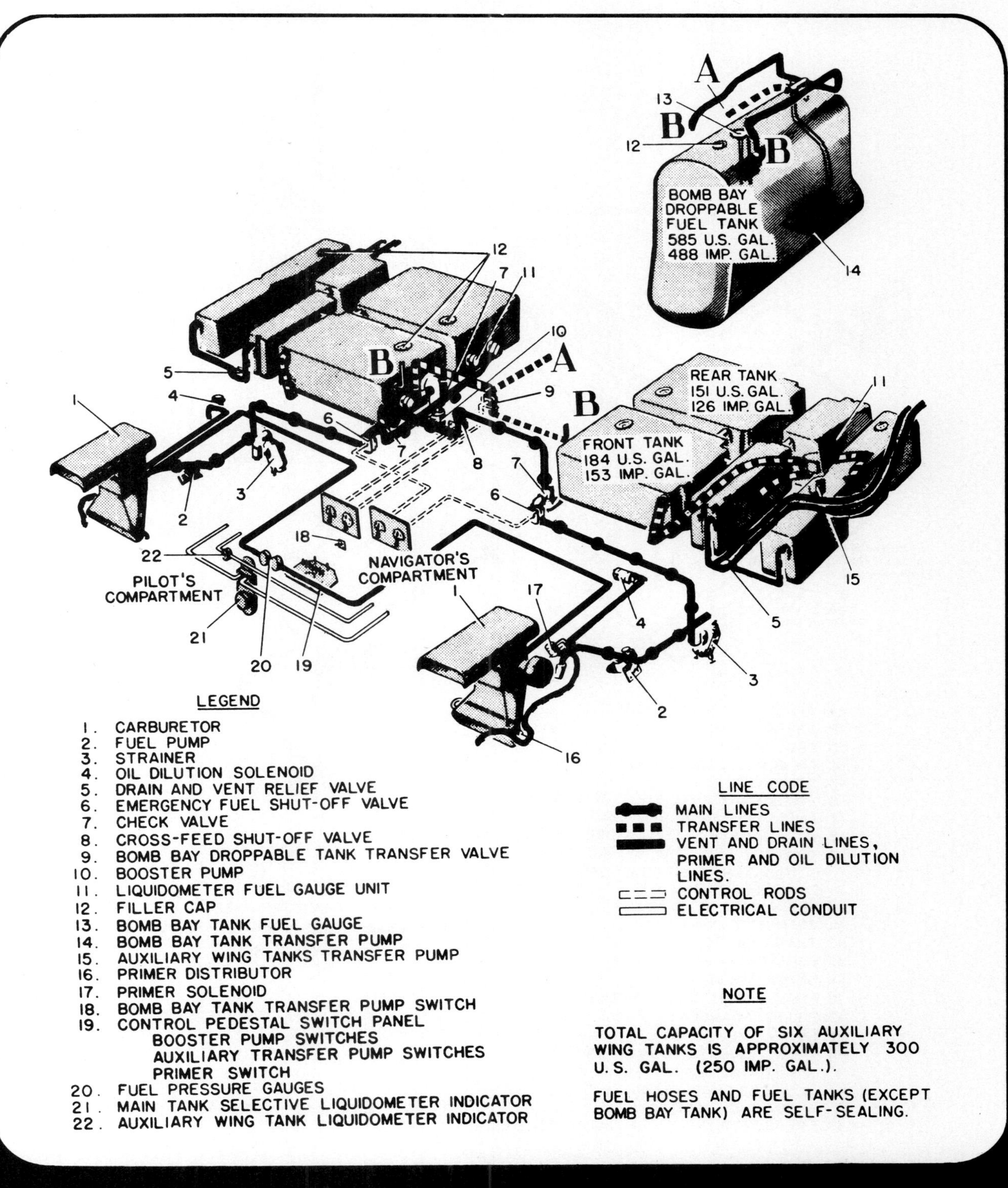

Figure 5 - Fuel System Diagram (Applicable to B-25C Airplanes, Serial Nos. AC-41-12817 and Subsequent, and B-25D Airplanes, Serial Nos. AC-41-29848 and Subsequent)

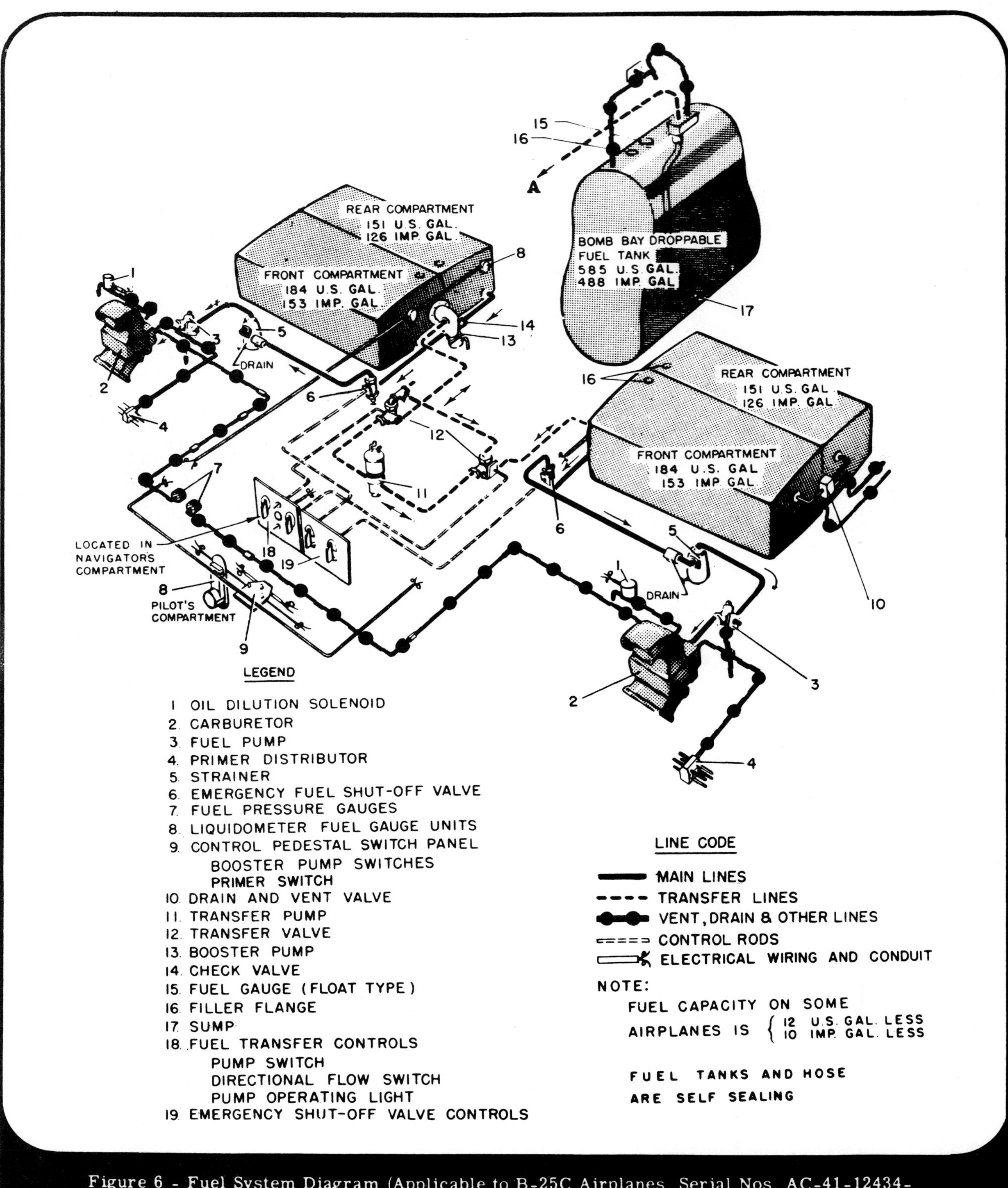

Figure 6 - Fuel System Diagram (Applicable to B-25C Airplanes, Serial Nos. AC-41-12434-12816 Inclusive, only, and B-25D Airplanes, Serial Nos. AC-41-29648-29847 Inclusive, only)

i. __Heating and Ventilating System.__

(1) A hot air heater is located in the left wing center section, and operates only on the left engine. (See figure 7.)

(2) The flow of cold or warm air is controlled by the pilot. The navigator regulates the temperature of the air. Controllable cold air scoops are provided for the pilot, copilot, and bombardier. Do not open bombardier's, pilot's, or side escape hatches for ventilation.

(3) __Defrosting System.__ - The pilot's windshield sections, the bomb sight window and bomb sight may be defrosted by warm air from the airplane heating system. Each outlet is controllable, except bomb sight window, which receives heat whenever heating system is on. A door is provided for cleaning bomb sight window. On airplanes equipped with the type A-3 automatic pilot, pilot's windshield also receives heat whenever system is on.

__k.__ __Oxygen Equipment.__ - Three low pressure G-1 oxygen bottles are installed in each nacelle. Eight type A-9 regulators (figure 8) are provided at the following stations:

Bombardier at bomb sight seat and riding seat.
Pilot on left side wall.
Copilot on right side wall.
Navigator on right side wall.
Radio operator on right side wall.
Upper turret operator on left side wall.
Camera operator on aft bulkhead.

__l.__ __Ice Eliminating Equipment.__ - De-icer shoes for the control surfaces and anti-icer slinger rings for the propellers are provided. (See figures 9 and 10.) The fluid capacity of the anti-icer tank is 4.5 U.S. gallons (3.8 Imperial gallons). Tank is located in rear of navigator's compartment. Fill with 85 percent denatured alcohol and 15 percent glycerine. When the de-icer shoes are not in operation, a vacuum pump prevents raising of the de-icer shoes. A pressure gage for the de-icer system is located in the rear of the navigator's compartment.

__m.__ __Protective Armor.__ - Armor plate is provided on the seat and back of the bombardier's seat. Armor plate is attached to the rear of pilot's and copilot's seats. An armor plate bulkhead is provided aft of turret gun. The armor plate door should be kept closed in flight. (See figure 11).

2. __POWER PLANT.__

The B-25C and B-25D airplanes are equipped with two 14-cylinder, double row, two-speed supercharged R-2600-13 air-cooled engines.

ENGINE LIMITATIONS B-25C and D AIRPLANES

Take-off	Low Blower	High Blower
RPM	2600 (for one minute)	DO
MP	44	NOT
Mixture	Full Rich	USE
Cowl Flaps and Oil Cooler Shutters	Open	
Cylinder Head Temperature		
(260° C (500° F) (5 minutes)		
Oil Temperature - 95° C (203° F)		

Climb	Below 11,000 ft	Above 11,000 ft
RPM	2400	2400
MP	38	39
Mixture	Full Rich	Full Rich
Cowl Flaps and Oil Cooler Shutters	Open	
Cylinder Head Temperature -		
260° C (500° F) (5 minutes)		
Oil Temperature - 95° C (203° F)		

High Speed	Below 11,000 ft	Above 11,000 ft
RPM	2400	2400
MP	38	39
Mixture	Full Rich	Full Rich
Cowl Flaps - Closed (Slightly open on warm days)		
Oil Cooler Shutters - Adjust to 60° C - 85° C		
(140° F - 185° F) Oil Temperature		
Cylinder Head Temperature - 218° C (428° F)		
Oil Temperature - 85° C (185° F)		

Cruising	Below 13,000 ft	Above 13,000 ft
RPM	2100	2100
MP	29.5 to 31.5 Max.	29.0 to 31.5 Max.
Mixture	Full Rich	Full Rich
Cylinder Head Temperature - 205° C (396° F)		

Cruising - Desired	Below 16,000 ft	Above 16,000 ft
RPM	2000	2000
MP	27	27
Mixture	Full Rich	Full Rich
Cylinder Head Temperature - 205° C (396° F)		

Cruising - Long Range	Below 13,000 ft	Above 13,000 ft
RPM	1550	1550
MP	27.5	26.5
Mixture	Cruising Lean	Cruising Lean
Cylinder Head Temperature - 205° C (396° F)		

Oil Pressure - 80° C to 90° C (176° F - 194° F)   Idle 25 Minimum

Fuel Pressure - 6-7 lb/sq in.

3. SUPERCHARGERS.

Two-speed turbo superchargers with a low blower ratio of 7.06:1 and a high blower ratio of 10.06:1 are provided. Type AF regulators are used.

4. PROPELLERS.

Two three-blade, hydromatic, full-feathering propellers with a diameter of 12 ft 7 in. are provided. A reserve oil supply is always available for feathering the propellers regardless of the exhaustion of the engine oil supply. The propeller pitch settings are 23 degrees and 90 degrees high.

5. OIL SYSTEM.

Each engine is provided with an independent oil system consisting of an oil tank, two automatic temperature regulators, an engine-driven pump, an oil dilution system, and pressure and temperature indicators. B-25C airplanes, serial Nos. AC-41-12817 and higher, and B-25D airplanes, serial Nos. AC-41-29848 and higher, are provided with self-sealing oil tanks and supply lines. Each oil compartment has a capacity of 37.5 U. S. gallons (31.2 Imperial gallons). (See figure 12.)

6. EMERGENCY EQUIPMENT. (See figures 46, 47, and 48, section X, Emergency Operations and Instructions.)

7. AUTOMATIC FLIGHT CONTROL SYSTEMS.

a. B - 25C Airplane. - Automatic flight control equipment is installed on all B-25C airplanes except serial Nos. AC-41-12457, -12459, -12461, -12463, -12465, -12467, -12469, -12471, -12473, -12475, -12477, -12479, -12517 and higher, which have the type A-3 automatic pilot installed.

b. B- 25D Airplane. - Automatic flight control equipment is installed on all B-25D airplanes except serial Nos. AC-41-29848 and higher, which have the type A-3 automatic pilot installed.

8. MOVEMENT OF FLIGHT PERSONNEL. - (See figure 48, section X, Emergency Operations and Instructions.)

9. MOORING.

a. Block main landing gear wheels, engage parking brakes, lock surface controls, and install hydraulic control locking plate. Also see that nose wheel tow pin is engaged.

b. Install the four wing mooring shackles, the one in the fuselage and the one on the underside of the tail skid. Type D-1 mooring kit will be used. (See figure 15.)

10. FLYING CHARACTERISTICS.

a. The general flying characteristics of the B-25C and B-25D airplanes are conventional for bi-motored bombardment airplanes.

b. Taxying Instructions.

(1) While taxying, the airplane should be turned by a gentle use of the brakes and engines, to avoid pivoting on one wheel. The minimum radius of turn of the inside wheel can be approximately ten feet (3 meters).

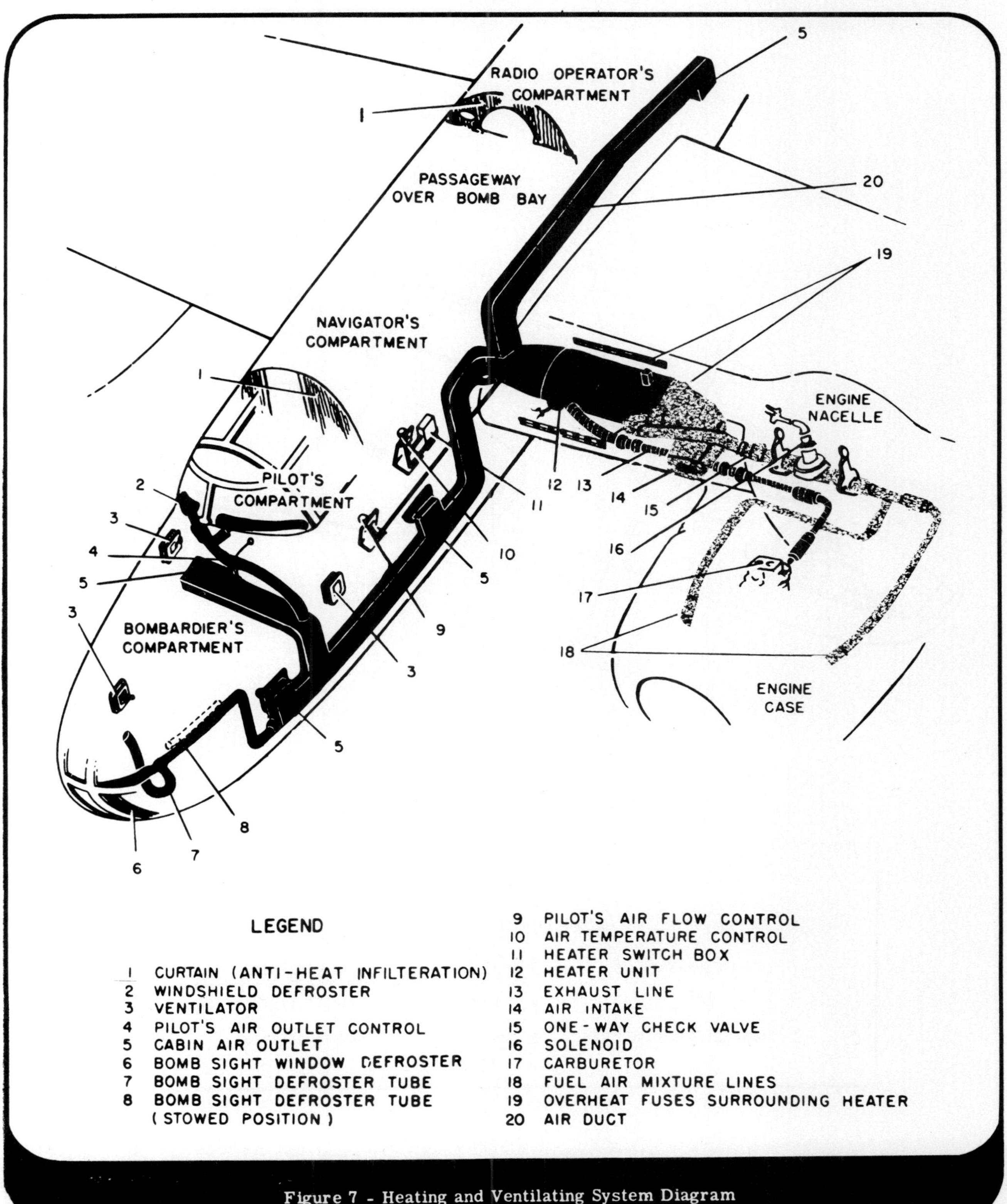

LEGEND

1 CURTAIN (ANTI-HEAT INFILTERATION)
2 WINDSHIELD DEFROSTER
3 VENTILATOR
4 PILOT'S AIR OUTLET CONTROL
5 CABIN AIR OUTLET
6 BOMB SIGHT WINDOW DEFROSTER
7 BOMB SIGHT DEFROSTER TUBE
8 BOMB SIGHT DEFROSTER TUBE (STOWED POSITION)
9 PILOT'S AIR FLOW CONTROL
10 AIR TEMPERATURE CONTROL
11 HEATER SWITCH BOX
12 HEATER UNIT
13 EXHAUST LINE
14 AIR INTAKE
15 ONE-WAY CHECK VALVE
16 SOLENOID
17 CARBURETOR
18 FUEL AIR MIXTURE LINES
19 OVERHEAT FUSES SURROUNDING HEATER
20 AIR DUCT

Figure 7 - Heating and Ventilating System Diagram

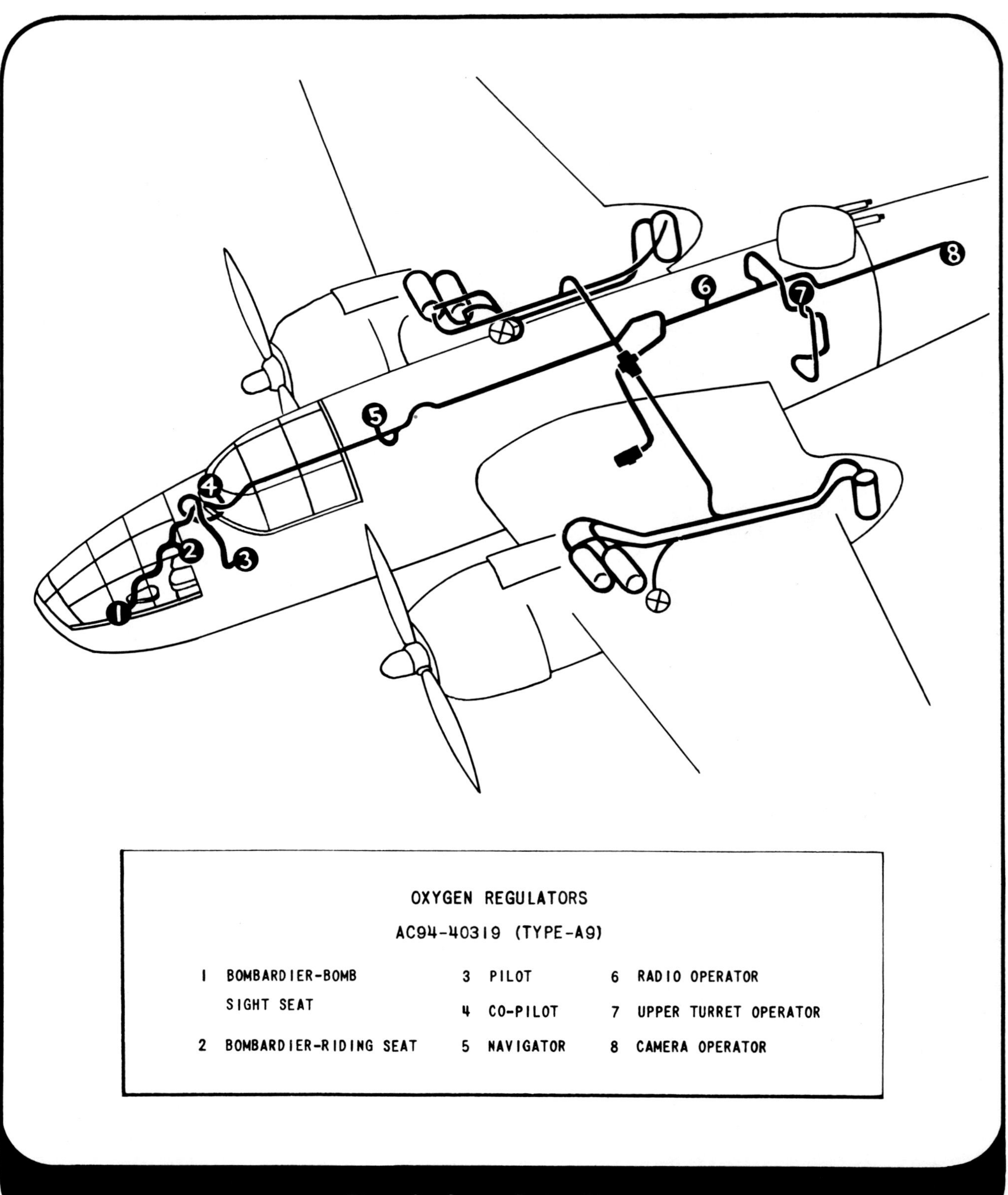

Figure 8 - Oxygen System Diagram

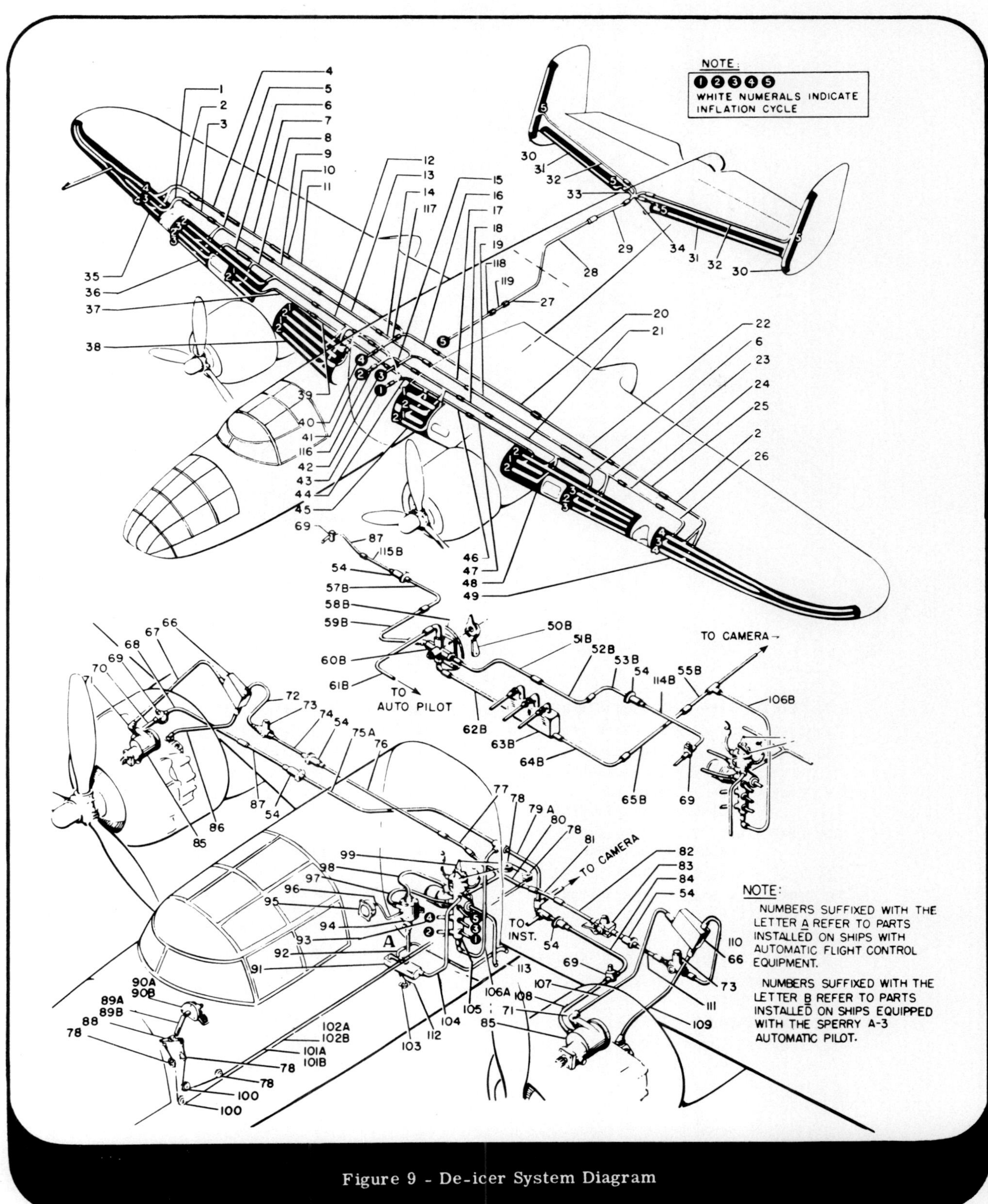

Figure 9 - De-icer System Diagram

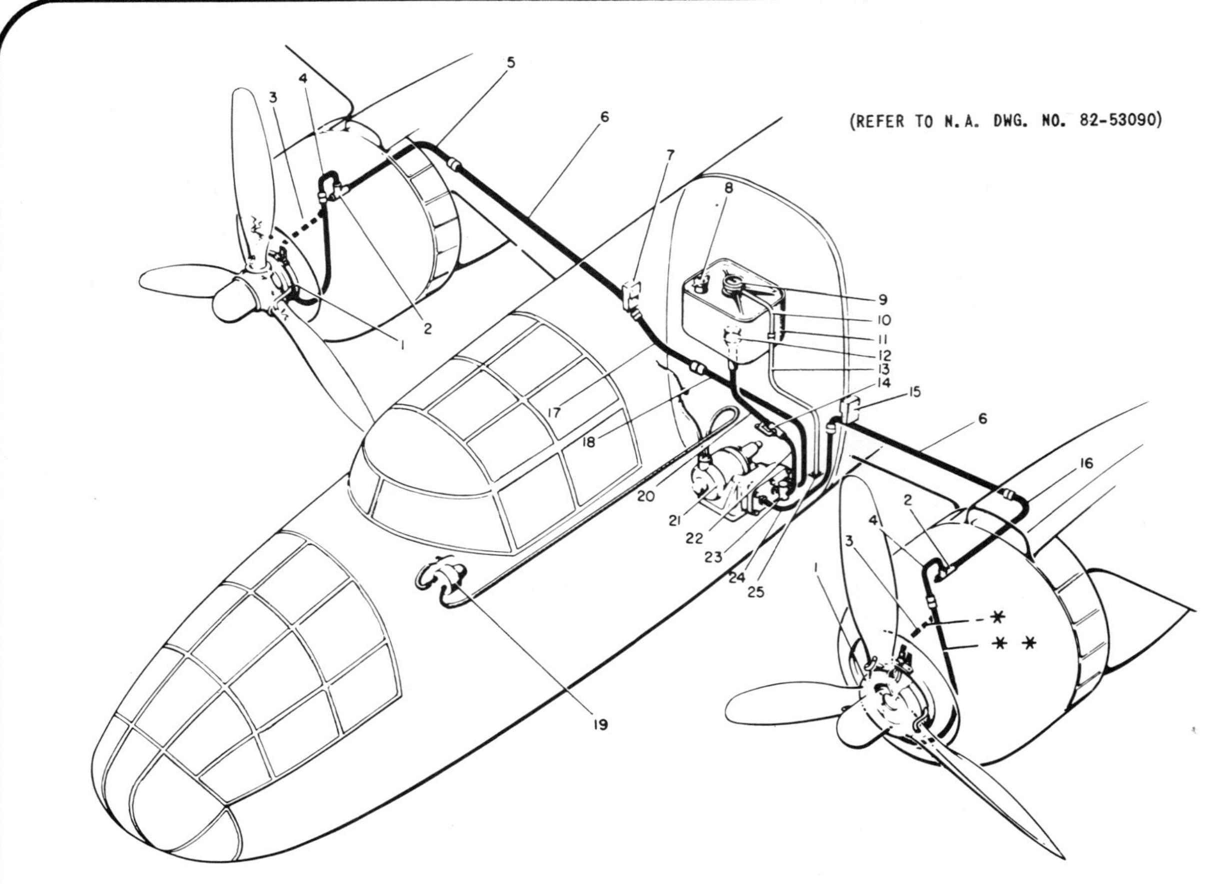

1. SLINGER RING (HAMILTON STANDARD)
2. CHECK VALVE - ANTI-DRAIN (PARKER)
3. LINE - NACELLE CHECK VALVE TO SLINGER RING 2ND SECTION
4. LINE - NACELLE CHECK VALVE TO SLINGER RING 1ST SECTION
5. LINE - WING STA. 96 TO NACELLE CHECK VALVE RIGHT
6. LINE - WING STA. 27-96
7. SUPPORT - WING CENTERSECTION LINE LEFT
8. CAP - RESERVOIR FILLER
9. FLANGE - FLUID GAGE DIAL RETAINING
10. LINE ASSEMBLY - TANK DRAIN 1ST SECTION
11. TANK ASSEMBLY - SUPPLY
12. FLOAT ASSEMBLY - SUPPLY TANK FLUID GAGE
13. LINE - TANK DRAIN 2ND SECTION
14. VALVE - SHUT-OFF (PARKER)
15. SUPPORT ASSEMBLY WING CENTER-SECTION LINE RIGHT
16. LINE ASSEMBLY WING STATION 96 TO NACELLE CHECK VALVE LEFT
17. LINE - UNION TO WING BLOCK RIGHT
18. LINE - FLUID PUMP TO UNION RIGHT
19. RHEOSTAT, ANTI-ICER (OHMITE)
20. LINE - TANK TO VALVE
21. PUMP (ECLIPSE)
22. LINE - VALVE TO FLUID PUMP
23. FILTER (ECLIPSE)
24. LINE - FLUID PUMP TO UNION LEFT
25. DRAIN - SUPPLY TANK FLUID

Figure 10 - Anti-icer System Diagram

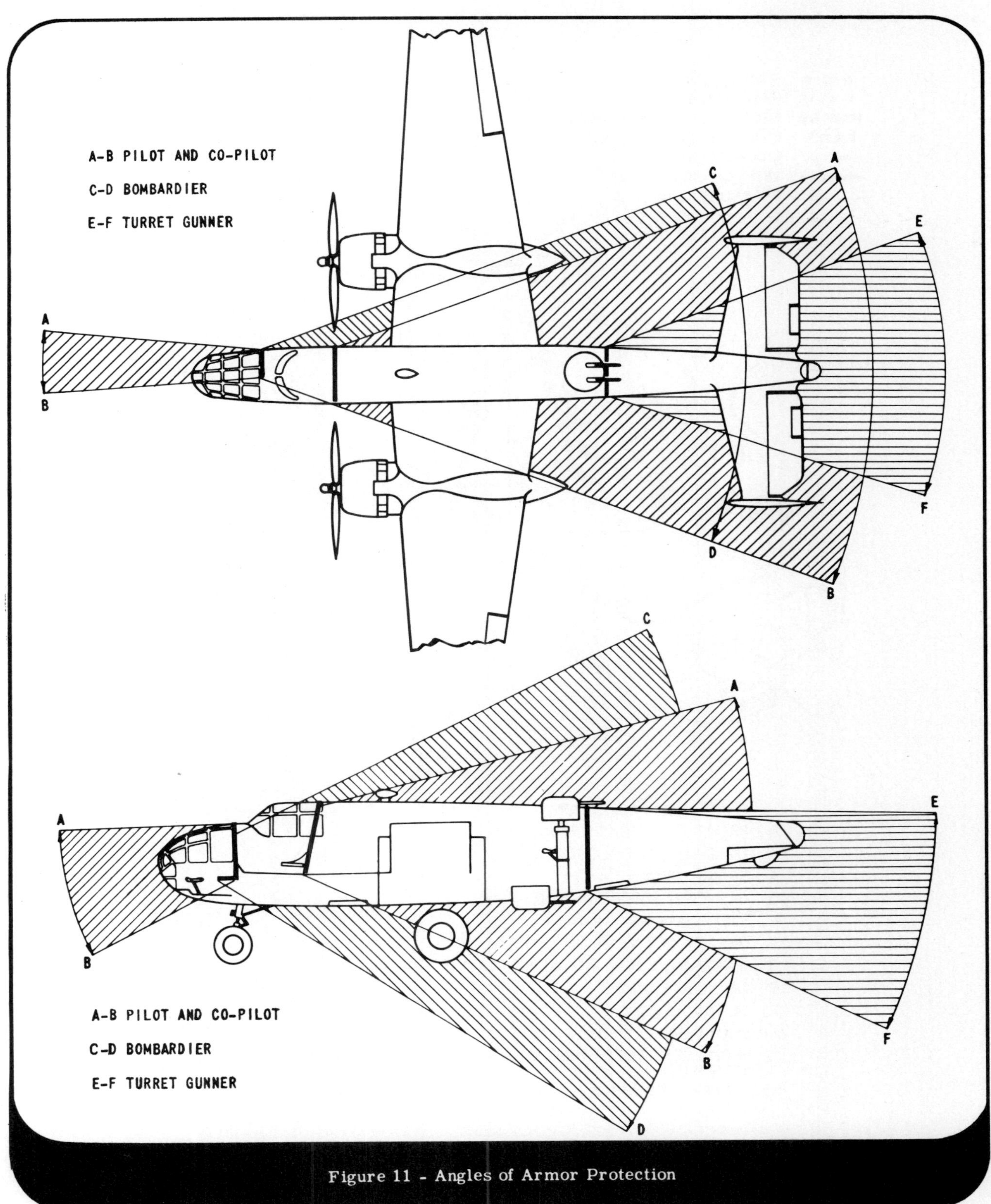

Figure 11 - Angles of Armor Protection

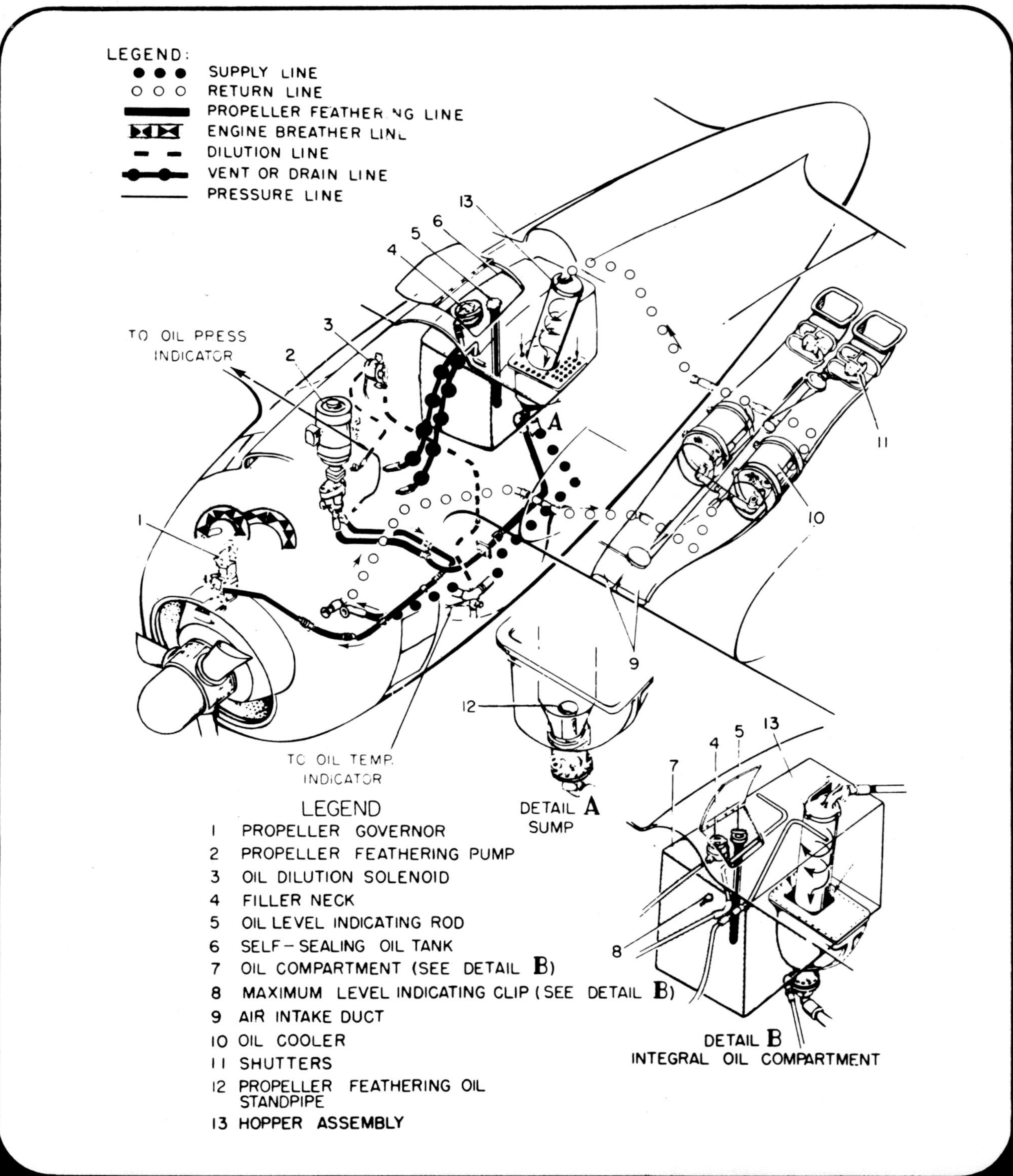

Figure 12 – Oil System Diagram

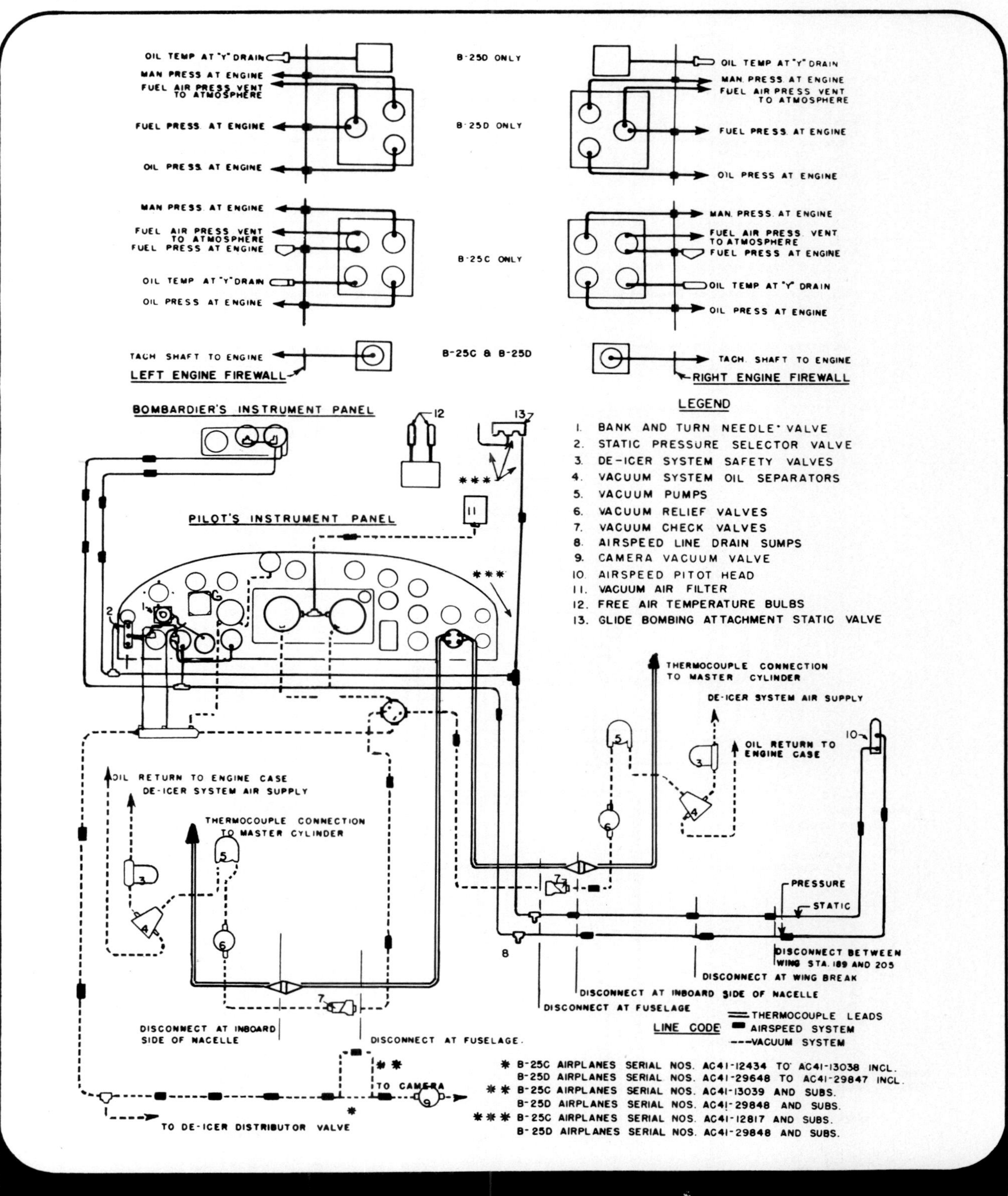

Figure 13 - Instrument and Vacuum System Diagram (A-3 Sperry)

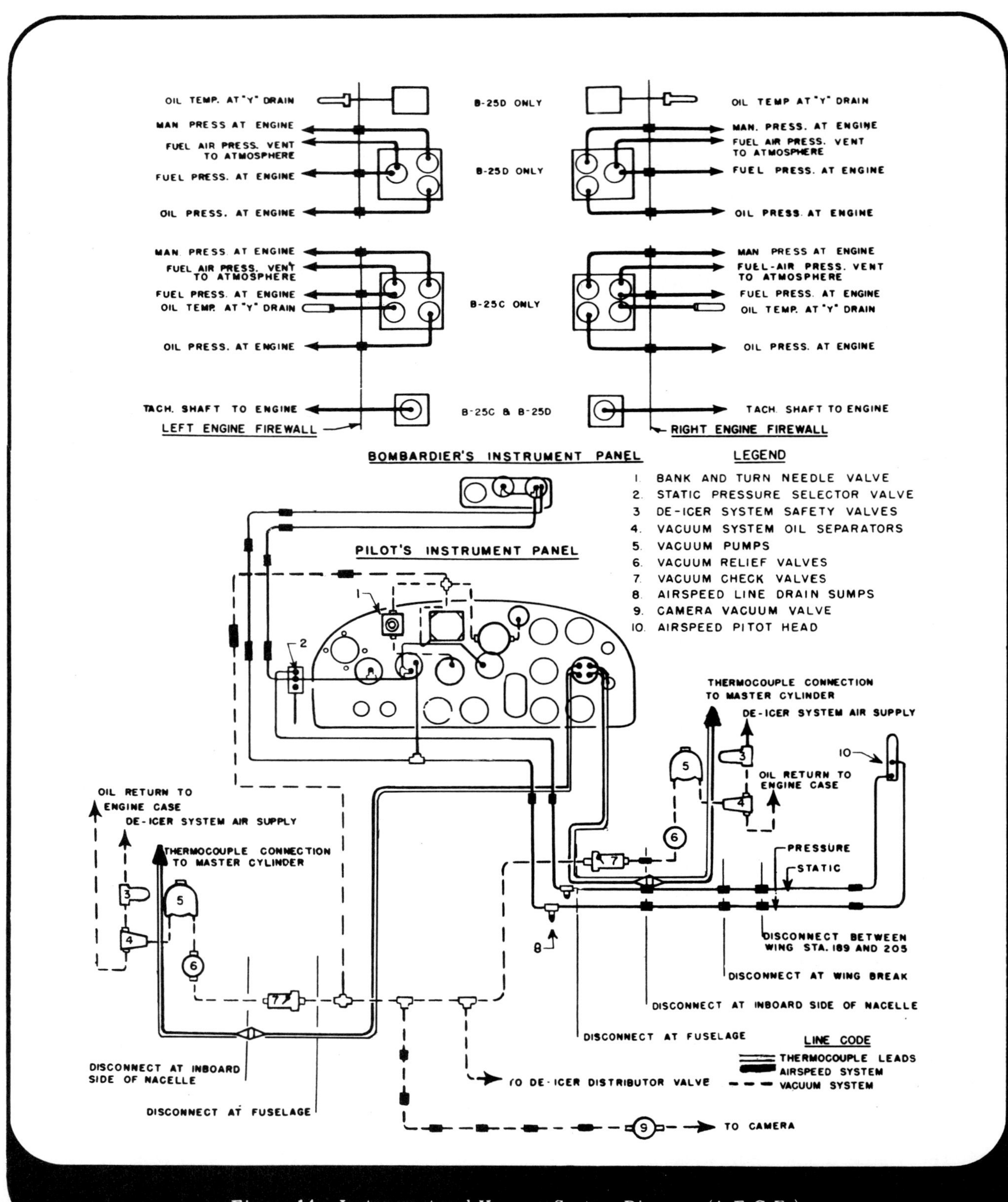

Figure 14 - Instrument and Vacuum System Diagram (A.F.C.E.)

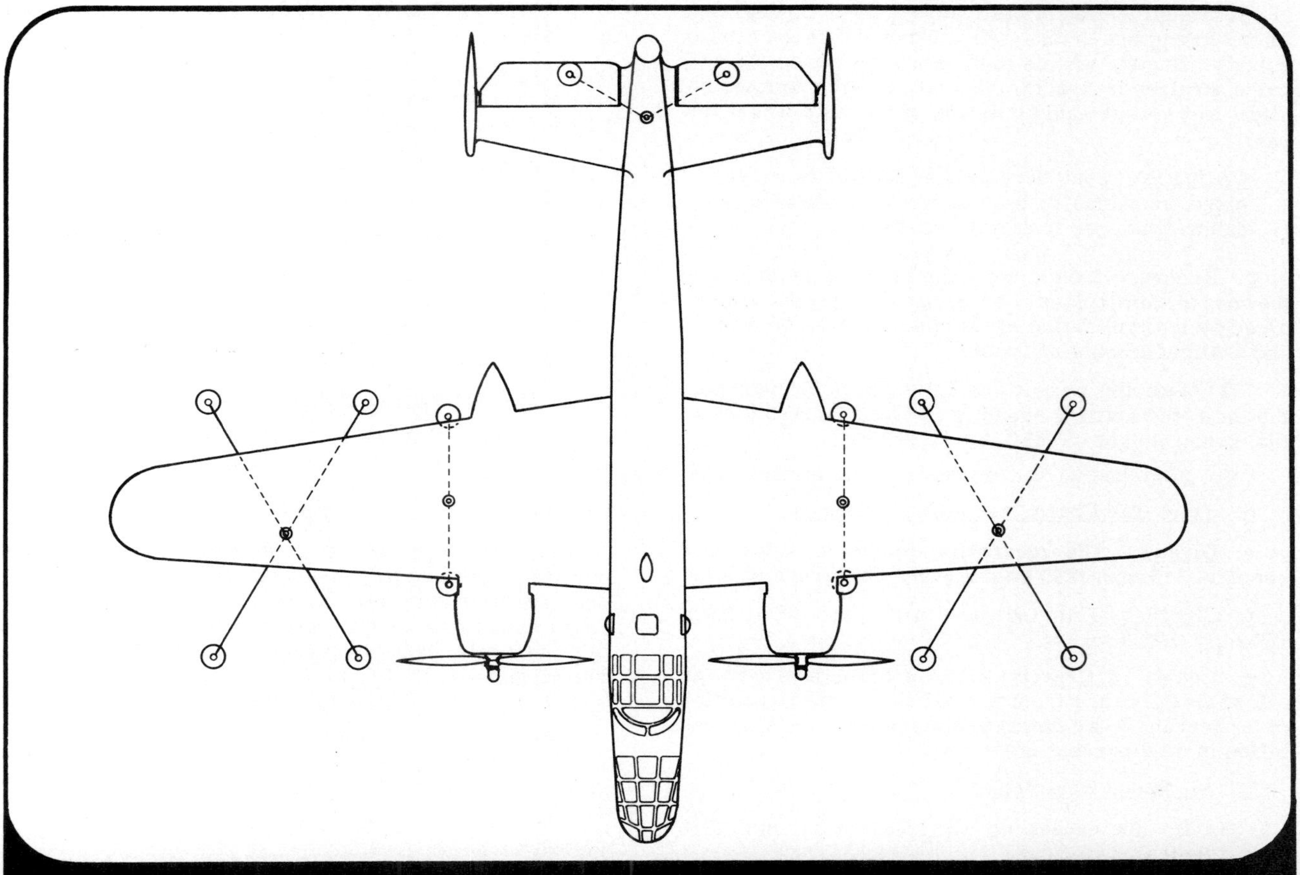

Figure 15 - Mooring Diagram

(2) When attempting to straighten the airplane out of a turn it will be found that the nose wheel has less tendency to trail properly as the depth of the tire sink increases. A tire sink in excess of 2 in. (5cm) causes a negative trailing reaction on the nose wheel, and, rather than straighten out, it will attempt to reverse its position and trail 180 degrees to its normal angle. If the tire sink exceeds 2 in. (5cm), one member of the crew should walk ahead of the airplane and signal to the pilot the position of the nose wheel and advise whether the nose wheel tends to trail properly or is attempting to reverse its position. If the airplane is pivoted on one wheel, the nose wheel will assume a sharp angle toward the rear, and the stationary pivot wheel will mire itself.

(3) If the airplane becomes mired:

(a) Manually turn nose wheel in line with the airplane fuselage by the use of a towing bar or a piece of iron pipe.

(b) If either of the main wheels sinks to a depth of 4 or 5 in. (10 or 12 cm), dig a shallow trench in front of each wheel with a forward slope of 10 to 20 degrees. Then run engines to take-off power and release brakes quickly. If main wheels sink more than 5 in. (12 cm) move airplane by tractor. Do not attempt to tow airplane over soft ground by means of the nose wheel tow bar.

> CAUTION: Considerable care should be exercised in manually turning the nose wheel because of danger from propellers.

c. Recovery from a stall can be made by dropping the nose of the airplane. A rolling tendency is counteracted by the application of the opposite aileron control and a slight amount of rudder.

(1) With the wing flaps fully down 45 degrees, the indicated stalling speed is reduced 20 mph at normal gross weight (25,450 lb; 11,542 kg).

(2) For stalling speeds refer to paragraph 10.m.

d. Trim Tab Changes Are Conventional.

e. Dives. - The maximum indicated diving air speed is 340 mph (295 knots) at normal gross weight.

f. Climbing. - An indicated air speed of at least 155 mph (133.3 knots).

g. Center of Gravity. - These airplanes have an allowable CG range from a maximum forward position of 20 percent (gear down) to a maximum rearward position of 32.6 percent (gear up).

h. Air Speed Restrictions.

(1) Do not exceed an air speed of 340 mph (295 knots) indicated.

(2) Do not lower landing gear at a speed in excess of 170 mph (148 knots) indicated.

(3) With landing gear down and wing flaps up do not exceed an air speed of 200 mph (174 knots) indicated.

(4) Do not lower main landing gear or nose gear by hand (mechanically) at a speed over 150 mph (130 knots) indicated.

(5) Do not lower wing flaps or fly with wing flaps down over 170 mph (148 knots) indicated.

(6) Do not lower wing flaps by hand or fly after the flaps are mechanically operated, at indicated air speeds over 150 mph (130 knots).

> CAUTION: Do not attempt to operate the wing flap mechanically or even install crank except when a complete failure of the hydraulic system exists, since the hydraulic and mechanical systems oppose each other. For complete instructions refer to section X.

(7) Do not open bomb bay doors at an air speed over 290 mph (252 knots) indicated.

(8) Do not operate de-icer system at speeds above 230 mph (200 knots) indicated.

(9) Do not operate de-icer system during take-off or landing.

i. Single Engine Operation. - The minimum controllable speed with one engine is 140 mph (122 knots). Fly with dead engine high.

j. Power Plant Restrictions.

(1) Do not exceed an engine speed of 2880 rpm.

(2) Do not idle engines on ground below 450 rpm.

(3) Do not turn on high blower during take-off.

> CAUTION: Use air pressure brake system in an emergency only. After using, hydraulic brake system must be bled.

k. Restricted Use of Automatic Pilot.

(1) Do not operate airplane by automatic pilot in extremely turbulent air, when de-icer system is operating, or when one or more engines are not delivering normal power output.

(2) Do not place airplane under control of the automatic pilot at any speed or altitude until the pilot has determined by manual operation that the existing flight conditions permit safe control by the automatic pilot, and in no case will automatic pilots be used when the airplane is flying at less than an indicated air speed of 40 mph (35 knots) above the stalling speed.

(3) Do not use automatic pilot unless one rated pilot remains "on watch" and maintains a close check of the airplane and instruments.

(4) Do not engage automatic pilot when follow-up indexes are not lined up.

(5) Do not make course and altitude changes by rapid knob movements. Turn slowly and smoothly.

(6) Do not allow airplane to get too far out of trim.

(7) Do not turn any of the three speed controls to "OFF" or lowest speed when automatic pilot is engaged, as this would lock the corresponding surface controls in whatever position they happen to be.

> CAUTION: Automatic pilot can be overpowered.

l. Loading Limitations. - Only straight flying is permitted when airplane is loaded to specified maximum loaded weight for safe flight. Refer to paragraph p. for maneuvers prohibited.

m. Stalls. - The stalling speeds for these airplanes with flaps up are as follows:

Level flight	101 mph;	87.7 knots
30 degree bank	134 mph;	116.3 knots
60 degree bank	154 mph;	133.7 knots

(1) Recovery from a stall may be made by dropping the nose.

(2) Stalling characteristics are not affected by changes of the gross weight, amount of power used, the setting of the wing flaps, or the operation of the

de-icer shoes. However the indicated stalling speed is raised 4 mph by operation of the de-icer. With the wing flaps fully down (45 degrees) the indicated stalling speed is reduced 20 mph at normal gross weight (25,450 lb; 11,542 kg.). A slight buffeting on the elevator and the horizontal stabilizer warns of a stall.

n. Spins. - Intentional spins are prohibited. However if an unintentional spin develops recovery may be made in the conventional manner.

CAUTION: These airplanes are very clean aerodynamically and will accelerate very rapidly in a dive. Recovery from any dive position should be without delay and accomplished as smoothly as possible.

o. Acrobatics. - All acrobatics are prohibited.

p. Maneuvers Prohibited.

Loop	Inverted Flight
Spin	Vertical Bank
Roll	Immelman

q. Diving.

(1) The maximum indicated air speed is 340 mph (295 knots) at normal gross weight.

(2) Recovery from dives should not be too abrupt.

(3) The following table gives maximum accelerations and diving speeds according to basic weight conditions:

Gross Weight	lb.	20,000	23,000	26,000	29,302	32,000	34,000
	kg.	9,070	10,430	11,791	13,291	14,512	15,419
Maximum allowable pull-out and push-over acceleration "g"		+3.67	+3.67	+3.67	+2.67	+2.67	+2.67
		-2.00	-2.00	-2.00	-1.67	-1.67	-1.67
Maximum allowable during speed (indicated)	mph	340	340	340	340	332	303
	knots	295	295	295	295	288	263

SECTION II

PILOT'S OPERATING INSTRUCTIONS

| Pilot | Copilot |

1. BEFORE ENTERING THE PILOT'S COMPARTMENT.

a. Check that nose gear towing pin is engaged.

b. To enter front entrance hatch:

(1) Verify that the lock on the front entrance hatch is unlocked.

(2) Depress flush type spring latch by pushing upward.

(3) Enter navigator's compartment.

c. While still in navigator's compartment:

(1) Check fuel cross-feed shut-off valve. (See figure 23-221.)

(2) Check fuel transfer valves. (See figure 23-222.)

<table>
<tr><td align="center">Pilot</td><td align="center">Copilot</td></tr>
</table>

Pilot

(3) Check generator main line switches. (See figure 23-224.)

(4) Check generator voltage switch. (See figure 23-225.)

(5) Check emergency nose gear operating pawl control (figure 25-239) "OFF."

2. ON ENTERING THE PILOT'S COMPARTMENT

a. Check for all flights.

(1) Check with radio operator that the main landing gear and wing flap emergency cranks <u>are</u> stowed.

(2) Check with navigator that bomb door emergency cranks <u>are</u> stowed.

(4) Check with crew members to see that the three emergency ground escape hatches are unlocked.

(19) Phone navigator to set emergency fuel shutoff valves to "ON."

Copilot

(3) Check emergency brake air pressure at 400-425 lb/sq in. (28.1 to 29.9 kg/sq cm)

(5) Ignition switches (figure 18-156) "OFF."

(6) Set parking brakes. (See figure 17-153.)

(7) Unlock surface controls by pulling up on lock handle (figure 18-161) and check operation.

(8) Remove snap wire hook attached to floor below landing gear control lever from the control handle (figure 21-192) and stow.

(9) Remove locking plate from wing and cowl flap control levers (figure 17-149) and stow the lock.

(10) See that automatic pilot or automatic flight control equipment is "OFF."

(11) If engines are inoperative for over two hours have propellers turned by hand three or four revolutions with the ignition "OFF."

(12) Crack open throttle 1/2 inch (1.27 cm) at (700-800 rpm).

(13) Set propeller control (figure 18-165) to "INCREASE RPM."

(14) Set mixture control (figure 18-167) to "FULL RICH."

(15) Lock supercharger control (figure 17-146) at "LOW."

(16) See that oil cooler shutter controls (figure 17-151) are "CLOSED."

(17) Set carburetor air (figure 16-141) to "NORMAL."

(18) Set cowl flaps (figure 17-149) to "OPEN."

Figure 16 - Pilot's Instruments and Control Panel

93 Fluorescent Lights Switch
94 Landing Gear Warning Horn Release Switch
95 Propeller Anti-Icer Rheostat
96 Formation Lights
97 Clock
98 De-Icer Valve
99 Spring Latch for Side Window
101 Air-Speed Tube Static Pressure Selector Valve (Alternate Source)
102 Emergency Bomb Release
103 Pilot's Direction Indicator
104 Altimeter
105 Free Air Temperature
106 Air Speed
107 Vacuum Adjustment Knob
108 Radio Compass
109 Turn and Bank Indicator
110 Turn Indicator

111 Landing Gear and Wing Flap Indicator
112 Climb Indicator
113 Flight Control (Artificial Horizon)
114 Caging Knob
115 Vacuum Pressure
116 Fuel Selector Valve
117 Manifold Pressure
118 Fuel Pressure
119 Oil Temperature
120 Tachometer
121 Cylinder Temperature
122 Hydraulic Pressure
123 Hydraulic Brake Pressure
124 Oxygen Regulator
125 Oil Pressure
126 Passing Lights
127 Battery Disconnect (Left)
128 Battery Disconnect (Right)

129 Compass Light Rheostat
130 Pitot Heater
131 Dome Light Switch Panel
132 Lamp on Aileron Control
133 Cockpit Lights
134 Running Lights
135 Oil Dilution Both Engines
136 Identification Light Keying Switch
137 Identification Light Top and Bottom
138 Bombardier Signal
139 Locking Controls, Propeller, and Mixture
140 Locking Controls
141 Carburetor Air
142 Extension Light and Switch
143 Lamp on Aileron Control
144 Fire Extinguisher Selector Valve and Pull Handle

▶ Figure 17 - Pilot's Control Pedestal

145 Rudder Trim Tab (Altitude)
146 Supercharger Control
147 Supercharger Control
148 Wing Flaps
149 Cowl Flaps
150 Cowl Flaps
151 Oil Cooler Shutters
152 Oil Cooler Shutters
153 Parking Brake

Figure 18 - Pilot's A.F.C.E. Control Panel ◀

154 Propeller Feathering Knob
155 Aileron (A.F.C.E.)
156 Left Ignition Switch
157 Elevator (A.F.C.E.)
158 Right Ignition Switch
159 Rudder (A.F.C.E.)
160 Propeller Feathering Knob
161 Surface Control Lock Handle
162 Landing Lights
163 Fuel Booster Pump
164 Throttle
165 Propeller
166 Fuel Booster Pump
167 Mixture Control
168 Primer Switch
169 Mixture Control
170 Energizing Switch
171 Starter Switch

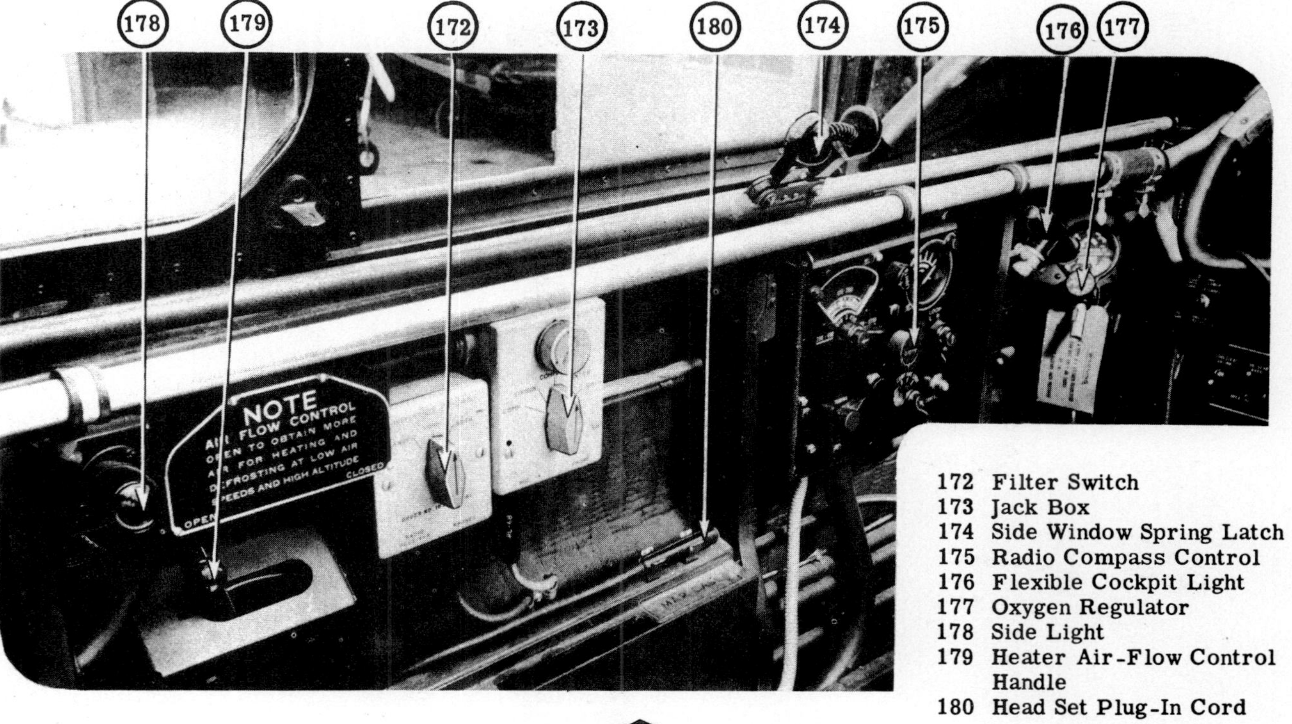

172 Filter Switch
173 Jack Box
174 Side Window Spring Latch
175 Radio Compass Control
176 Flexible Cockpit Light
177 Oxygen Regulator
178 Side Light
179 Heater Air-Flow Control Handle
180 Head Set Plug-In Cord

Figure 19 - Pilot's Compartment (Left Side)

Figure 20 - Pilot's Compartment (Right Side)

181 Oxygen Regulator
182 Flexible Cockpit Light
183 Air Scoop
184 Jack Box
185 Filter Switch
186 Side Window Spring Latch
187 Head Set Plug-In
188 Receiver
189 Transmitter
190 Side Light

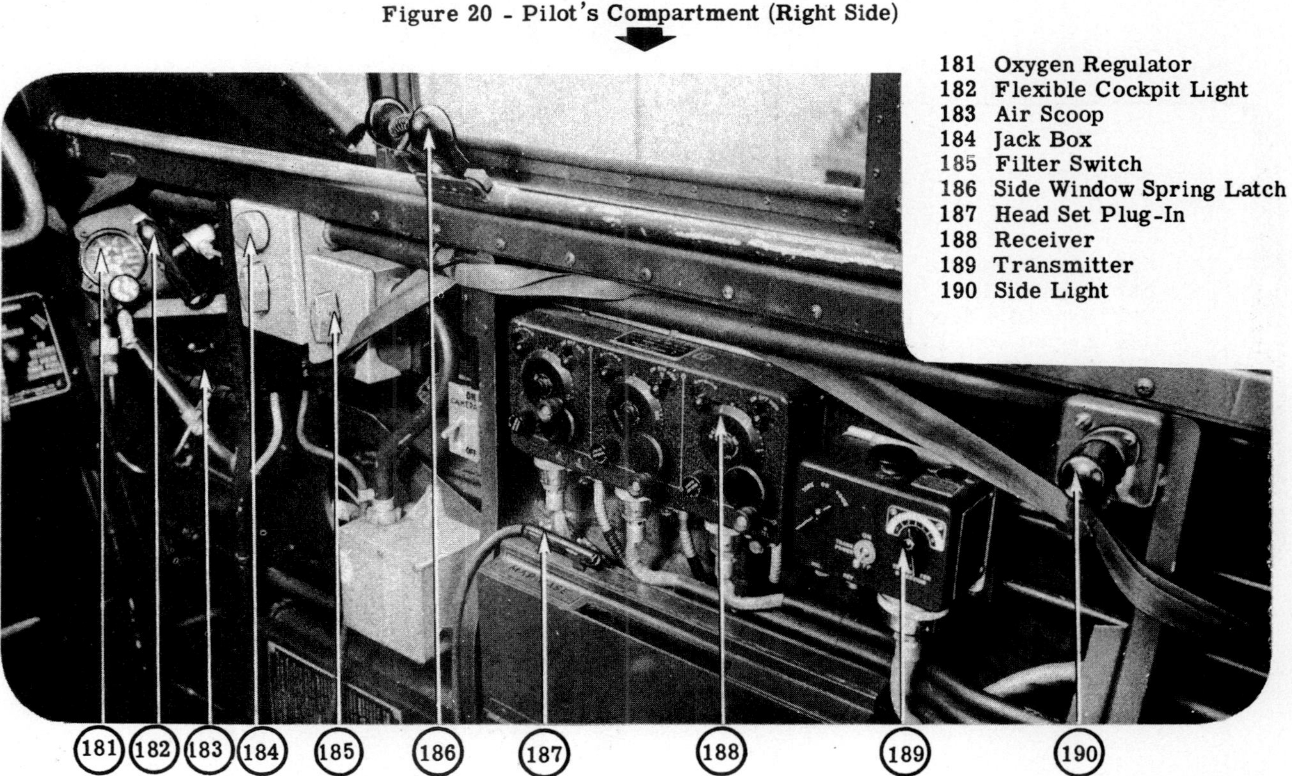

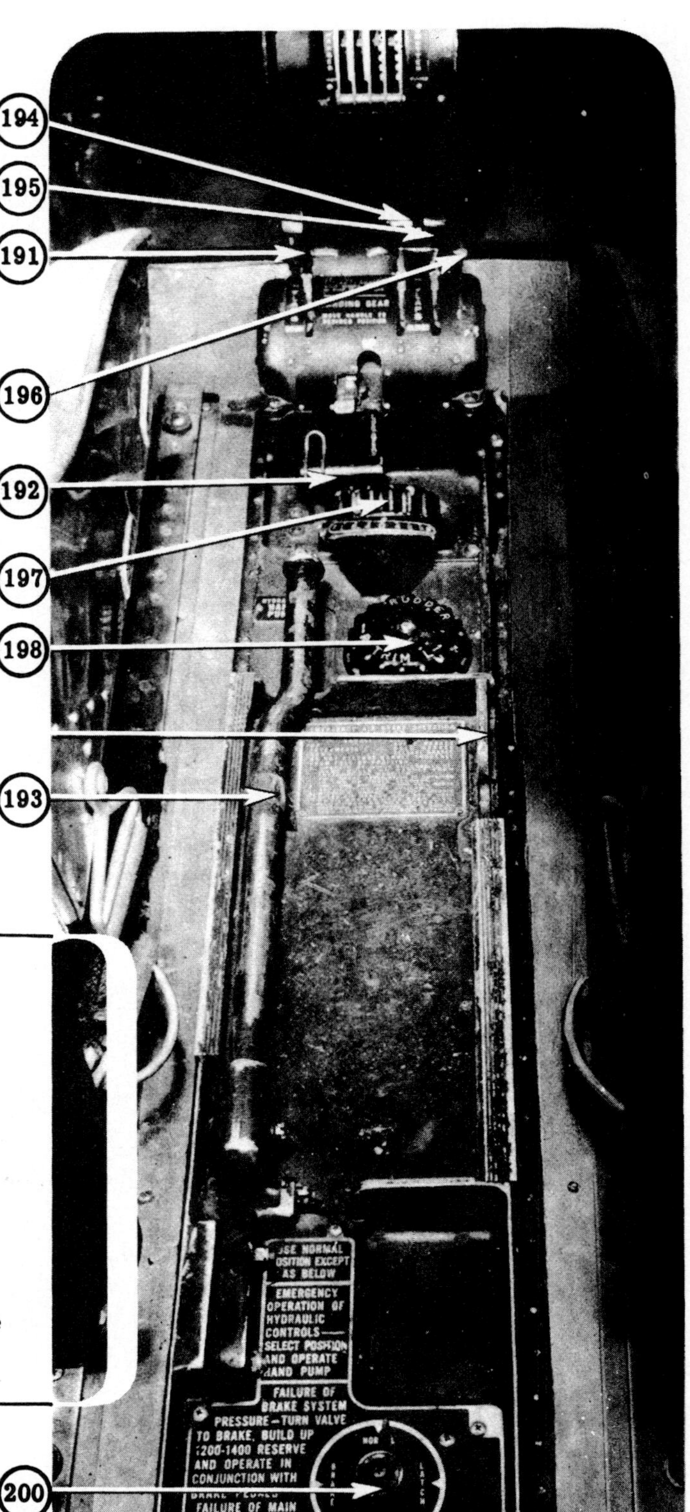

Figure 21 - Floor Controls Between
Pilot and Copilot

191 Wing Flaps
192 Landing Gear
193 Hydraulic Hand Pump
194 Engine Cowl Flaps
195 Engine Cowl Flaps
196 Control Lock
197 Aileron Trim
198 Rudder Trim
199 Emergency Air Brake
200 Emergency Hydraulic Selector Valve

<u>Pilot</u>   <u>Copilot</u>

3. STARTING ENGINE.

    <u>a</u>. Check with navigator to see that battery disconnect, generator-ignition safety, and active inverter switches are "ON."

    <u>b</u>. Set booster pump switch (figure 18-166) to "ON" (fuel pressure 6 to 7 lb/sq in. (.4 to .5 kg/sq cm).

    <u>c</u>. Turn on engine primer switch (figure 18-168); prime 3 to 5 seconds if cold; 2 seconds if hot.

    <u>d</u>. Set fire extinguisher switch (figure 16-144) to "ON."

    (1) If engine fails to start due to cold, instruct a crew member or a member of the ground crew to use hand crank. Then press engaging switch (figure 18-171) to "ON," followed by pressing energizing switch (figure 18-170) to "ON;" then hold both switches "ON."

    <u>e</u>. Prime while engaging as required.

4. ENGINE WARM-UP.

    <u>a</u>. As engine starts, check oil pressure gage (figure 16-125); if not 40 lb/sq in. (2.8 kg/sq cm) after 30 seconds, stop engine and investigate. Warm engine at 1200 rpm until oil temperature shows a definite increase and the oil pressure remains steady when throttle is opened.

    <u>b</u>. Set booster pump switch (figure 18-166) "OFF" (fuel pressure 6 to 7 lb/sq in. (.4 to .5 kg/sq cm).

    <u>c</u>. Open oil cooler shutter controls (figure 17-151) to oil temperature of 40° degrees C (104 degrees F).

    <u>d</u>. Check elevator trim tab control. (See figure 17-145.)

    <u>e</u>. Check aileron trim tab control. (See figure 21-197.)

    <u>f</u>. Check rudder trim tab control. (See figure 21-198.)

    <u>g</u>. See that wing flap control handle (figure 17-148) is in "CLOSE" position.

    <u>h</u>. Check each magneto at 1800-2000 rpm by momentarily turning each ignition switch off in turn; maximum manifold pressure 27 to 31.5 in. Hg (68.6 to 80 cm Hg)(maximum rpm drop 75). If drop is greater, run at same rpm for 15 seconds and recheck.

    <u>i</u>. At 1600 rpm have navigator check voltmeter (figure 23-225) at 28 to 28.5 and the ammeter (figure 23-223) 40 to 60 maximum.

5. EMERGENCY TAKE-OFF.

    <u>a</u>. If the engines were properly diluted (figure 16-135) when previously stopped, no trouble should be experienced in maintaining oil pres-

<u>Pilot</u>

sure of 80 to 90 lb/sq in. (5.6 to 6.3 kg/sq cm). Then proceed with normal take-off.

6. <u>ENGINE AND ACCESSORIES GROUND TEST.</u>

<u>b</u>. Check flight controls for free and proper movement, and watch control surfaces for correct response.

<u>c</u>. Check de-icer control (figure 16-98) "OFF."

<u>d</u>. Check with crew to see hatches are closed and that lower turret is retracted.

<u>e</u>. Check with navigator on fuel level gages.

<u>f</u>. Check suction (figure 16-115): 3.75 to 4.25 in. Hg (9.53 to 10.8 cm Hg).

<u>r</u>. Check with navigator to see that emergency fuel shut-off valves (figure 23-227) are "ON."

<u>s</u>. Check with navigator to verify heater switch (figure 22-212) "OFF."

7. <u>PREFLIGHT CHECK.</u>

<u>a</u>. Check flight controls for free and proper movement, and watch control surfaces for correct response.

<u>b</u>. Check for position of elevator, aileron and rudder trim tabs.

<u>c</u>. Wing flaps (figure 17-148) 20 degrees down (control neutral).

<u>Copilot</u>

<u>a</u>. See that automatic pilot or automatic flight control equipment (figure 16-114) is caged "OFF."

<u>g</u>. Check hydraulic pressure (figure 16-122) at 800 to 1100 lb/sq in. (56.2 to 77.3 kg/sq cm).

<u>h</u>. Check brake pressure (figure 16-123): 1000 to 1200 lb/sq in. (70.3 to 84.3 kg/sq cm).

<u>i</u>. Fuel booster pumps (figure 18-163) "ON." Fuel pressure 6 to 7 lb/sq in. (.4 to .5/sq cm).

<u>j</u>. Propeller controls (figure 18-165) at "INCREASE RPM" (locked snug).

<u>k</u>. Mixture controls (figure 18-167) at "FULL RICH" (locked snug).

<u>l</u>. Supercharger controls (figure 17-146) "LOW" (locked).

<u>m</u>. Oil cooler shutters (figure 17-151) "OPEN."

<u>n</u>. Carburetor air (figure 16-141) "NORMAL" or "ICING CONDITIONS" as required.

<u>o</u>. Cowl flaps (figure 17-149) "OPEN" (control neutral).

<u>p</u>. Emergency air brake (figure 21-199) safetied by wire in down position.

<u>q</u>. Emergency hydraulic selector valve (figure 21-200) "NORMAL."

<u>t</u>. Uncage gyro instruments. (See figure 16-114.)

<u>Pilot</u>

(1) For best obstacle clearance on short run set
flaps 30 degrees down.

8. <u>TAKE-OFF</u>.

<u>a</u>. Do not turn on heater during take-off.

<u>b</u>. Propeller controls (figure 18-165) increase rpm
(locked snug).

<u>i</u>. Refer to ''Take-off, Climb and Landing Chart.''

9. <u>ENGINE FAILURE DURING TAKE-OFF</u>.

<u>a</u>. Move mixture control (figure 18-169) to ''IDLE
CUT-OFF'' and cut the ignition switch of dead
engine.

<u>c</u>. Put nose of airplane down and make belly landing.

10. <u>CLIMB</u>.

<u>Copilot</u>

<u>c</u>. Engine rpm 2600 maximum and manifold pres-
sure (figure 16-117) 44 in. Hg (111.8 cm Hg).
Maximum time 1 minute (lock throttle snug).

<u>d</u>. Landing gear retracted only on definite signal
from pilot.

<u>e</u>. Fuel pressure (figure 16-118) 6 to 7 lb/sq in.
(.4 to .5 kg/sq cm).

<u>f</u>. Cylinder temperature (figure 16-121) 160 de-
grees C minimum to 260 degrees C maxi-
mum (320 degrees F minimum to 500 degrees
F maximum for 5 minutes).

<u>g</u>. Oil pressure (figure 16-125) 80 to 90 lb/sq in.
(5.6 to 6.3 kg/sq cm).

<u>h</u>. Oil temperature 40 degrees C (104 degrees F)
minimum 95 degrees C (203 degrees F) maximum.

<u>b</u>. If bomb bay droppable tank is installed, release
immediately if sufficient altitude has been at-
tained to open and close bomb doors. (See figure 16-102.)

<u>d</u>. Do not lower landing gear or nose wheel.

<u>e</u>. If bombs are being carried do not release.

<u>a</u>. Landing gear (figure 21-192) ''UP.''

<u>b</u>. Manifold pressure (figure 16-117) 38 in. Hg
(96.5 cm Hg); maximum below 11,000 ft (3350
meters). 39 in. Hg (99.1 cm Hg); maximum
above 11,000 ft (3350 meters).

<u>c</u>. Engine rpm 2400 maximum.

<u>d</u>. Fuel pressure (figure 16-118) 6 to 7 lb/sq in.
(.4 to .5 kg/sq cm).

<u>e</u>. Cylinder temperature (figure 16-121) 260 de-
grees C (500 degrees F) maximum for 15
minutes.

<u>f</u>. Oil pressure (figure 16-125) 80 to 90 lb/sq in.
(5.6 to 6.3 kg/sq cm).

<u>Pilot</u>

<u>i</u>. Carburetor air (figure 16-141) "NORMAL" or "ICING CONDITIONS" as required.

<u>j</u>. Supercharger: "LOW" below 11,000 ft (3350 meters); "HIGH" above 11,000 ft (3350 meters). Shift from low to high at 1400 to 2400 rpm.

<u>k</u>. Wing flaps (figure 17-148) "UP"; cowl flaps (figure 17-149) "OPEN" (controls neutral).

11. <u>FLIGHT OPERATION</u>.

<u>CAUTION</u>: Do not allow one fuel tank to run completely dry before switching to another tank.

(5) Mixture fuel (figure 18-167) "RICH."

(6) Supercharger "LOW" below 13,000 ft (4000 meters); "HIGH" above 13,000 ft (4000 meters). Shift from "LOW" to "HIGH" at 1400 to 2400 rpm.

(8) Carburetor air "NORMAL" or "ICING CONDITIONS" as required.

<u>NOTE</u>: Consult Automatic Pilot Check-Off list if airplane is so equipped.

12. <u>ENGINE FAILURE DURING FLIGHT</u>.

The minimum controllable speed with one engine at rated power is 140 mph, indicated. If the speed drops below 140 mph, the application of

<u>Copilot</u>

<u>g</u>. Oil cooler shutters (figure 17-151) "OPEN." Oil temperatures 95 degrees C (203 degrees F) maximum. (See figure 16-119.)

<u>h</u>. Mixture (figure 18-167) "FULL RICH."

<u>a</u>. Cruising.

(1) Engine rpm maximum and manifold pressure 31.5 in. Hg (80 cm Hg) maximum.

(2) Fuel pressure 6 to 7 lb/sq in. (.4 to .5 kg/sq cm). Fuel booster pumps "ON" as required to maintain 6 to 7 lb/sq in. (.4 to .5 kg/sq cm) fuel pressure.

(3) Oil pressure 80 to 90 lb/sq in. (5.6 to 6.3 kg/sq cm).

(4) Check suction 3.75 to 4.25 in. Hg (9.53 to 10.8 cm Hg).

(7) Adjust oil cooler shutters to obtain 60 degrees C to 85 degrees C (140 degrees F to 185 degrees F).

(9) Wing flaps and landing gear "UP" (flap control "NEUTRAL").

(10) Cowl flaps closed or open as required (control neutral). Cylinder temperature 205 degrees C (401 degrees F) maximum.

(11) Check volts 28 to 28.5; Amperes 40 to 60 maximum.

<u>Pilot</u>

full power on the single engine causes the airplane to yaw. Airplane should be dived at reduced power to attain proper speed. Adjust rudder trim tab to counteract loss of engine thrust and fly the airplane with the dead engine high.

13. <u>EMERGENCY CREW EXITS</u>.   (See section X.)

14. <u>APPROACH, LANDING, AND CROSS-WIND LANDING</u>.

<u>a</u>. <u>Landing</u>.

(2) Set de-icer control (figure 16-98) ''OFF.''

(4) Verify that lower gun turret is retracted.

(5) Check fuel levels with navigator.

(16) Check with navigator that heater control (figure 22-212) is ''OFF.''

<u>CAUTION</u>: If landing is not made, do not raise flaps until sufficient altitude and speed are obtained.

<u>b</u>. <u>Emergency Operation of Landing Gear and Flaps</u>.

(1) Order navigator to lower nose gear.

(2) At request of navigator, retard throttle momentarily to see that nose gear is locked

<u>Copilot</u>

(1) Turn automatic pilot (figure 16-114) ''OFF.''

(3) Cage gyro instruments. (Figure 16-114.)

(6) Check hydraulic pressure (figure 16-122) at 800 to 1000 lb/sq in. (52.6 to 77.3 kg/sq cm).

(7) Check brake pressure (figure 16-123) at 1000 to 1200 lb/sq in. (70.3 to 84.3 kg/sq cm).

(8) Check fuel booster pumps (figure 18-166) both ''ON.''

(9) Set propeller (figure 18-165) at 2100 rpm.

(10) Set mixture controls (figure 18-169) ''FULL RICH.''

(11) Lock supercharger control (figure 17-147) at ''LOW.''

(12) Oil cooler shutters (figure 17-151) ''OPEN.''

(13) Cowl flaps (figure 17-149) ''CLOSED'' (control neutral).

(14) Set landing gear (figure 21-192) ''DOWN'' and lock. Air speed should be less than 170 mph (148 knots) when lowering landing gear. Check operation by indicator (figure 16-111) and warning horn.

(15) Safety emergency brake control (figure 21-199) with pressure at 400 to 425 lb/sq in. (28.1 to 29.9 kg/sq cm).

(17) Set wing flaps (figure 17-148) ''DOWN'' (control neutral). Do not exceed 170 mph (148 knots).

<u>Pilot</u>

down, as evidenced by failure of warning horn to sound.

<u>c</u>. <u>Lowering of Main Landing Gear</u>.

(1) Order radio operator to lower main landing gear.

(2) Retard throttle momentarily to ascertain that the landing gear is locked down, as evidenced by the failure of warning horn to sound.

<u>d</u>. <u>Emergency Operation of Flaps</u>.

(1) Order radio operator to lower wing flaps.

(2) Move hydraulic flap control handle to the extreme position for the desired movement of the flaps.

<u>e</u>. <u>Emergency Take-Off If Landing Is Not Completed</u>.

(1) Full throttle.

(2) Leave landing gear and wing flaps down until sufficient altitude and speed are obtained.

15. STOPPING OF ENGINES.

16. BEFORE LEAVING PILOT'S COMPARTMENT.

<u>g</u>. If airplane is not to be serviced by a ground crew, lock all entrance hatches.

<u>Copilot</u>

<u>a</u>. Propeller at full "INCREASE RPM."

<u>b</u>. If necessary use oil dilution system.

(1) Hold the oil dilution switch (figure 16-135) "ON" for approximately four minutes while engines are idling at 800 rpm.

(2) Stop engines conventionally keeping the dilution switch on until the engines stop turning.

<u>c</u>. Idle at 600 to 800 rpm for one to three minutes until cylinder temperature is below 190 degrees C (374 degrees F).

<u>d</u>. Run engines at 1500 to 1600 rpm for five seconds, then move mixture control to "IDLE CUT-OFF" and open throttle. Do not move mixture control from "IDLE CUT-OFF."

<u>e</u>. After engine stops turn all switches "OFF."

<u>a</u>. Fuel booster pumps "OFF."

<u>b</u>. Propeller at full "INCREASE RPM."

<u>c</u>. Wing flaps "UP."

<u>d</u>. Cowl flaps "OPEN."

<u>e</u>. Lock flying controls.

<u>f</u>. Replace locks on hydraulic controls.

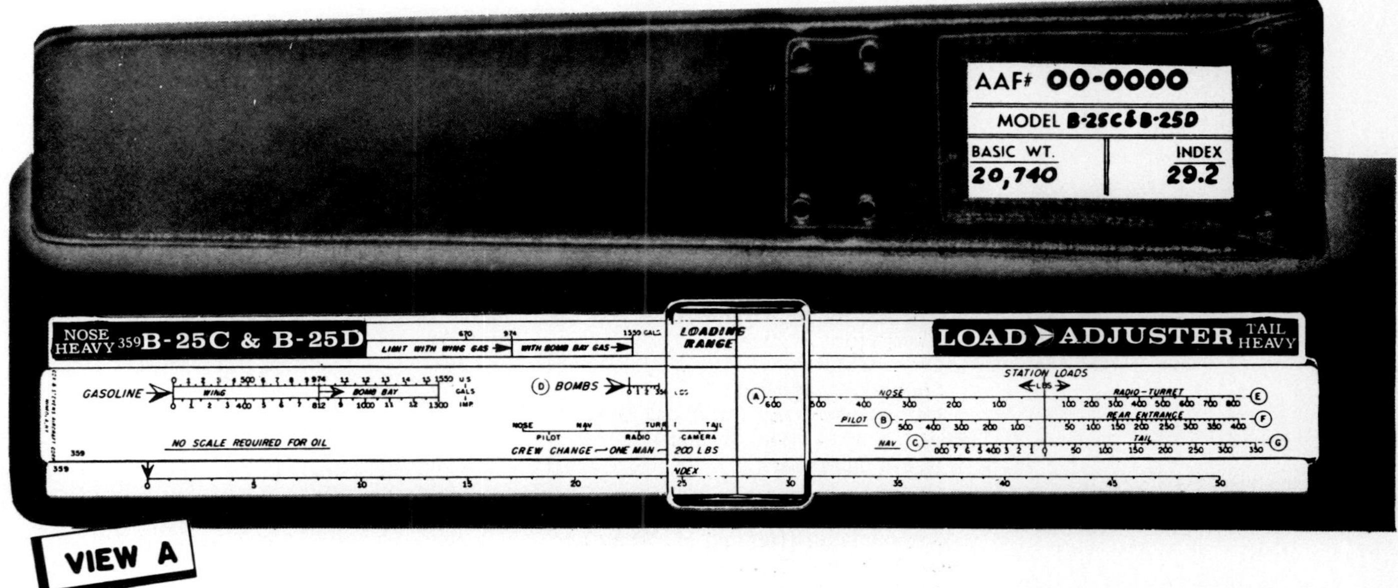

17. Instructions for Use of Load Adjuster.

a. Airplane Loading. - Check and sign the weight and balance clearance prepared by the ground loading personnel. This may be rapidly and accurately accomplished by using a load adjuster (view A). The following instructions and sample loading problem are published as condensed instructions for the information and guidance of all personnel using a load adjuster to determine change in balance from the basic airplane to the loaded airplane as flown, and to insure that the weight distribution of all items loaded above and beyond the basic airplane weight and balance will not produce a weight and balance condition beyond permissible limits.

b. Application of Load Adjuster. - A load adjuster and carrying case for the models B-25C and B-25D airplanes will be found located on a mounting hook adjacent to the data case. Pick up the instrument and ascertain that the serial number for the airplane being loaded is identical with the serial number inscribed on the carrying case identification card (view A).

CAUTION: The airplane model designation stamped on every load adjuster indicates that the instrument may be used for balance calculations on any AAF airplane of that particular model. However, the index figure entered on the carrying case identification card is correct only for the specific airplane serial number printed directly above, and represents the balance moment of only that one individual basic airplane.

c. Operating Instructions.

(1) The following sample loading problem is itemized in detail, and complete instructions with supporting illustrations are published to furnish the Service with complete instructions on loading aircraft above and beyond the basic airplane (including personal items and all items commonly referred to as "expendable" items) are to be taken into consideration for each and every loading problem, and their balance moments must be added with the load adjuster on the compartment scale where they are to be located.

(a) Given:

Item	Sub-Total	Total
Basic Airplane		20,740
Gasoline (974 U.S. gal - 812 Imp. gal)		6,744
Oil (42 U.S. gal - 35 Imp. gal)		315
Bombs (Bomb Bay)		3,500
Nose Compartment		525
Bombardier	200	
Special Equipment	300	
Handbook Data	25	
Pilots' Compartment		450
Brief Cases (2)	50	
Pilots (2)	400	
Navigator's Compartment		650
Navigator	200	
Crew Chief	200	
Special Equipment	200	
Navigational and Handbook Data	50	
Radio-Turret Compartment		800
Radio Operator	200	
Gunner	200	
Special Equipment	100	
Ammunition (1000 rd of .50 cal)	300	
Rear Entrance Compartment		400
Photographic Equipment and Supplies	125	
Photographer	200	
Brief Cases (3)	75	
Tail Compartment		125
GROSS WEIGHT		34,249

(b) **To Find:** If the load distribution brings the airplane balance within permissible cg limits as indicated on the load adjuster "loading range" scale.

(2) Set indicator hairline on basic airplane index 29.2 (obtain from identification card on the load adjuster carrying case), and move slide to the zero mark on the "GASOLINE" scale as shown in view B.

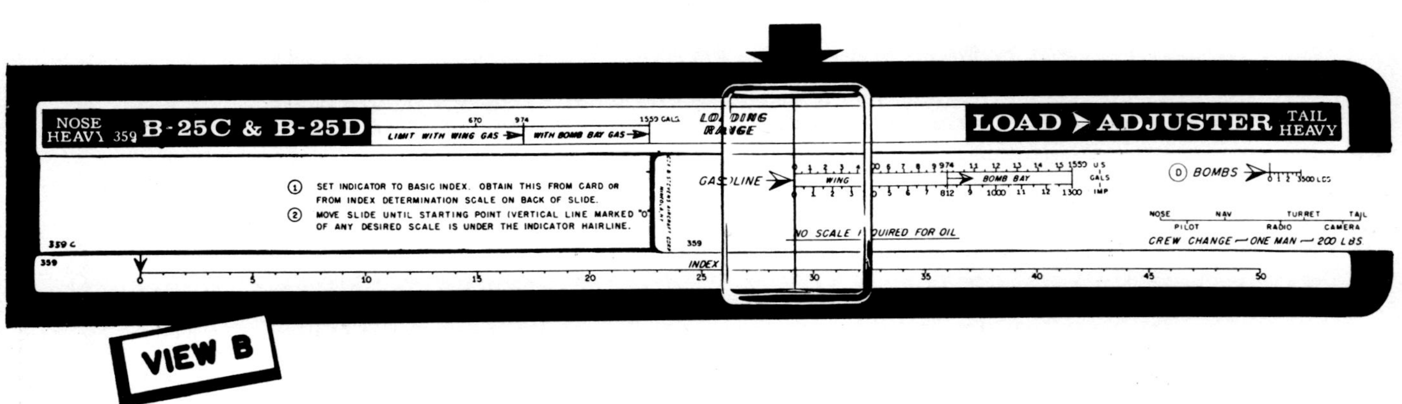

Move indicator until the hairline is over 974 on the top (U.S.) edge of the
scale. This adds the balance moment of 974 gallons of gasoline as loaded
in the airplane's wing tanks and moves the index to 36.0 as shown in view C.

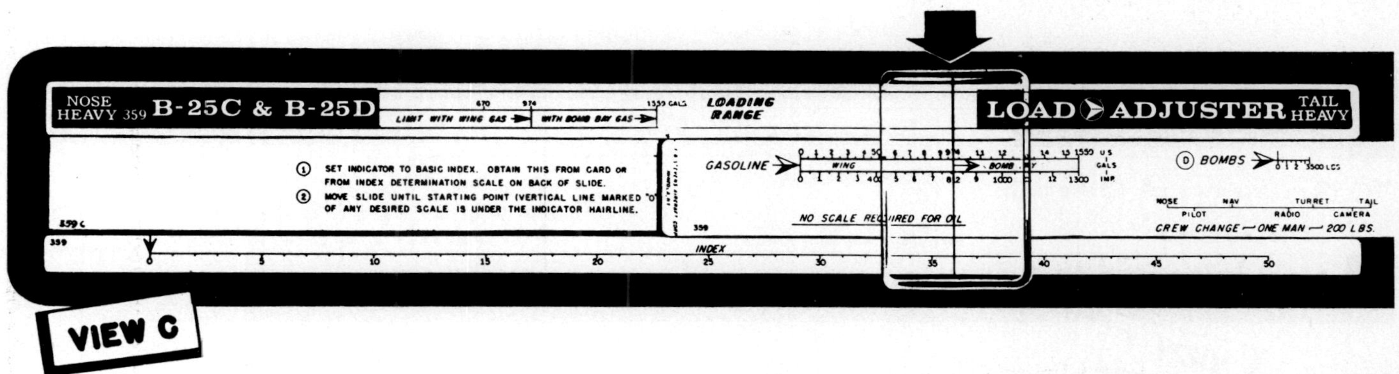

VIEW C

(3) No oil scale is necessary as the oil tank is located adjacent to the
best cg.

(4) Set slide to the zero mark on the "BOMBS" scale as illustrated
in view D.

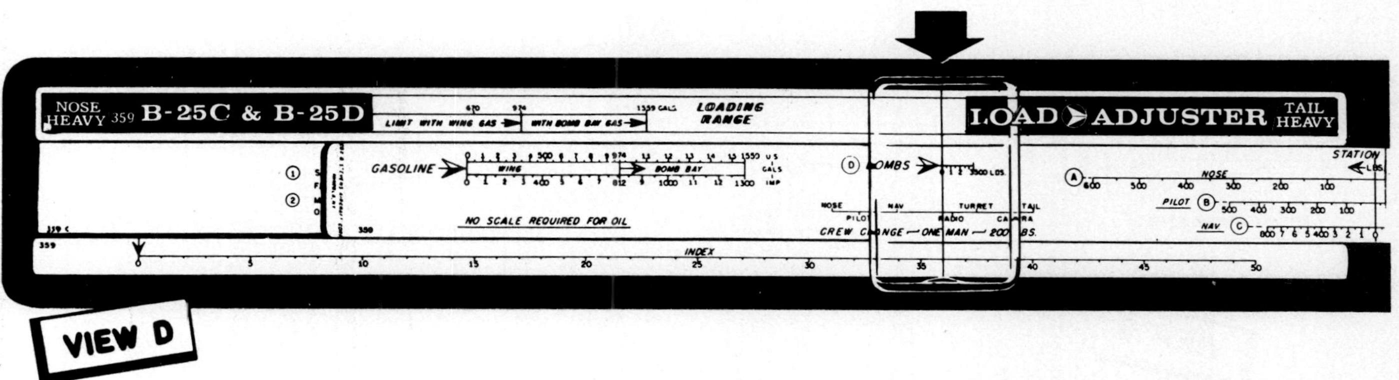

VIEW D

Move indicator until the hairline is over 3500 on the scale. This adds the
balance moment of 3500 pounds of bombs as loaded in the bomb bay and
moves the index to 37.4 as illustrated in view E.

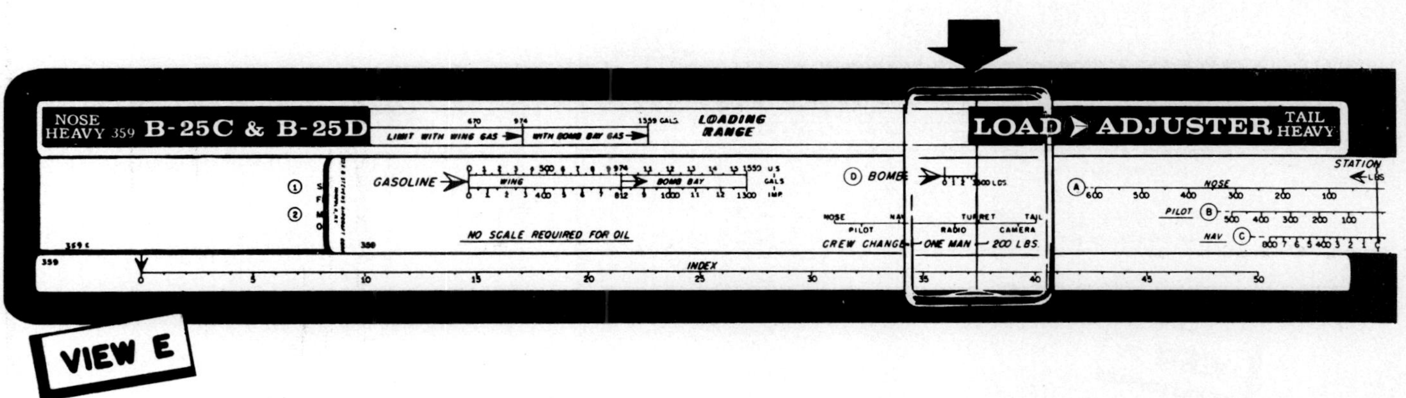

VIEW E

(5) Set slide to the compartment ("STATION LOADS") zero mark as illustrated in view F.

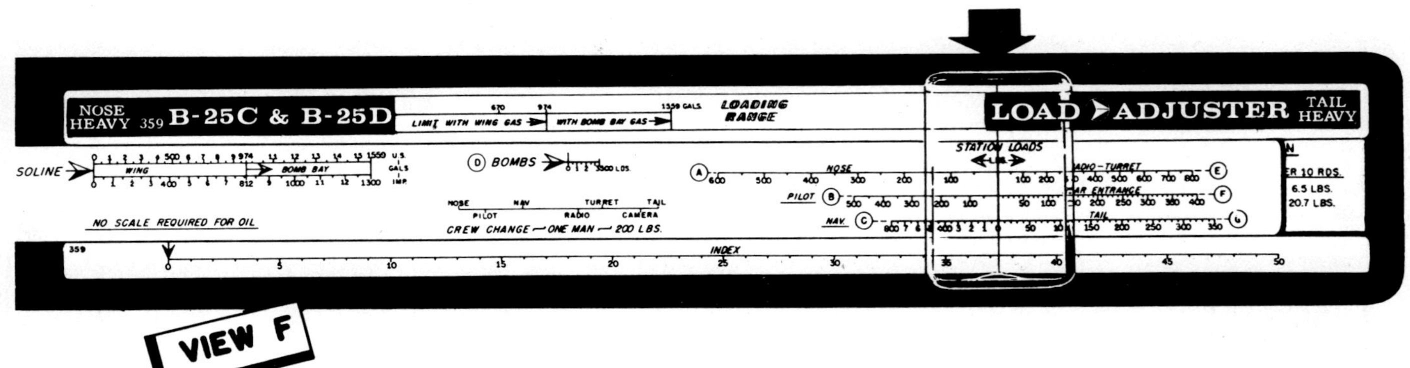

Move indicator until the hairline is over 525 on the "NOSE" scale. This adds the balance moment of the items that were loaded in that compartment and moves the index to 26.4 as illustrated in view G.

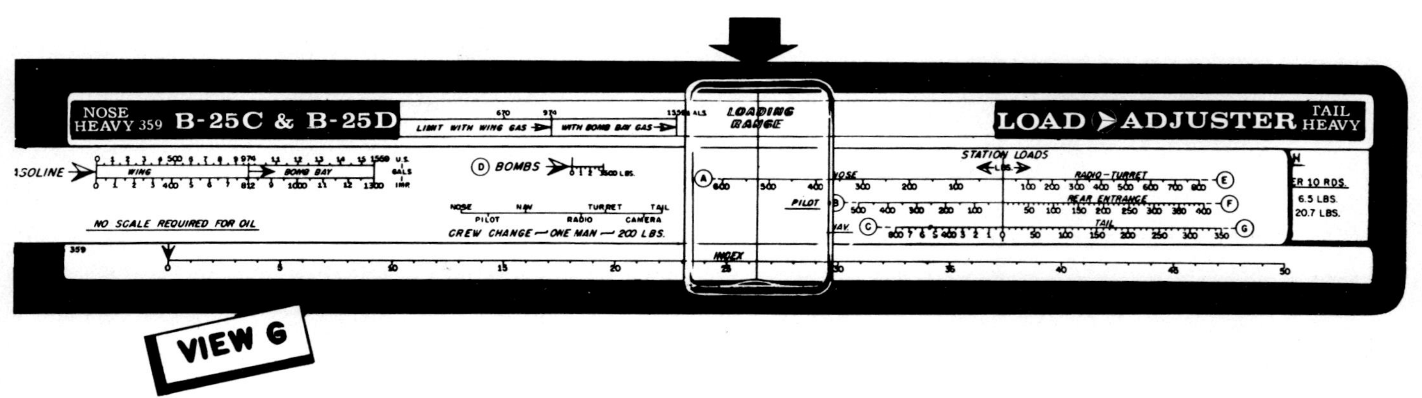

(6) Set slide to the compartment zero line as illustrated in view H.

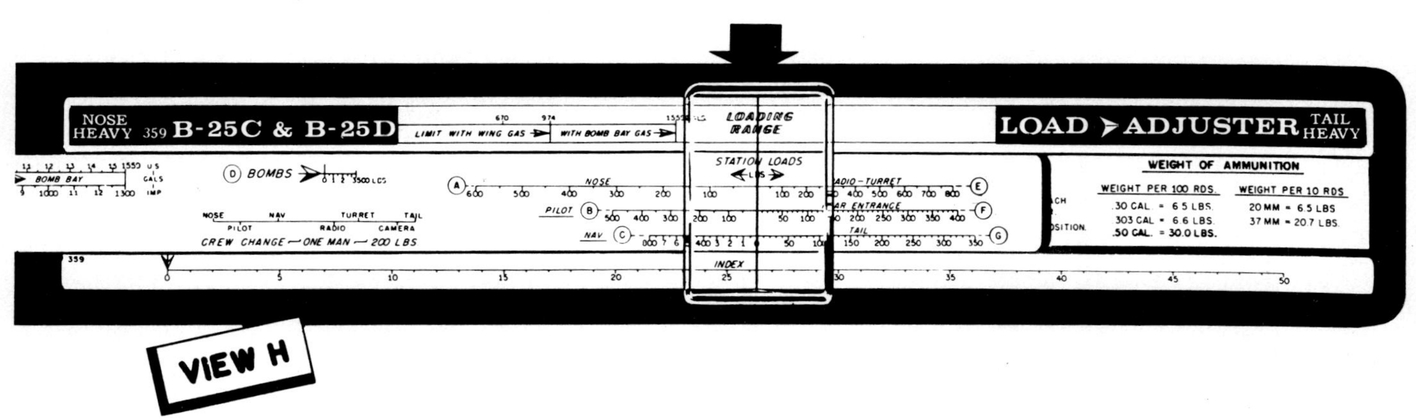

Move indicator until the hairline is over 450 on the "PILOT" scale. This adds the balance moment of the two pilots and their brief cases in the airplane and moves the index to 20.6 as illustrated in view I.

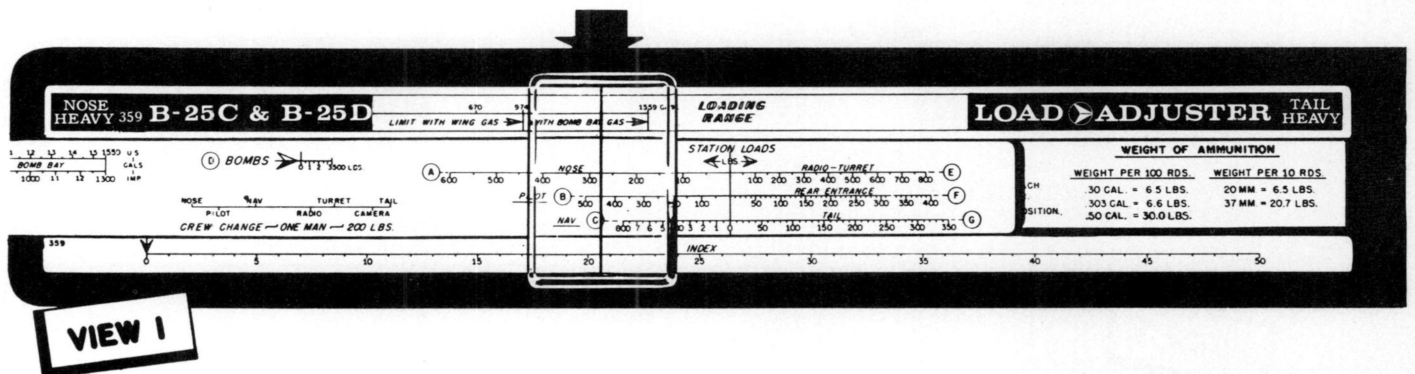

(7) Set slide to the compartment zero line as shown in view J.

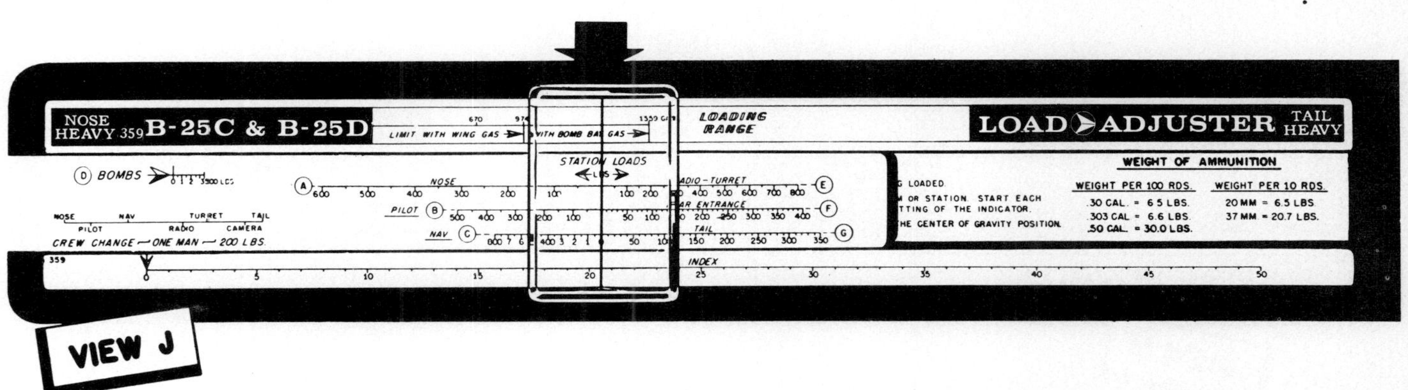

Move indicator until the hairline is over 650 on the "NAV." scale. This adds the balance moment of two men, 200 pounds of special equipment and 50 pounds of navigational and Handbook data as loaded in the airplane and moves the index to 16.6 as shown in view K.

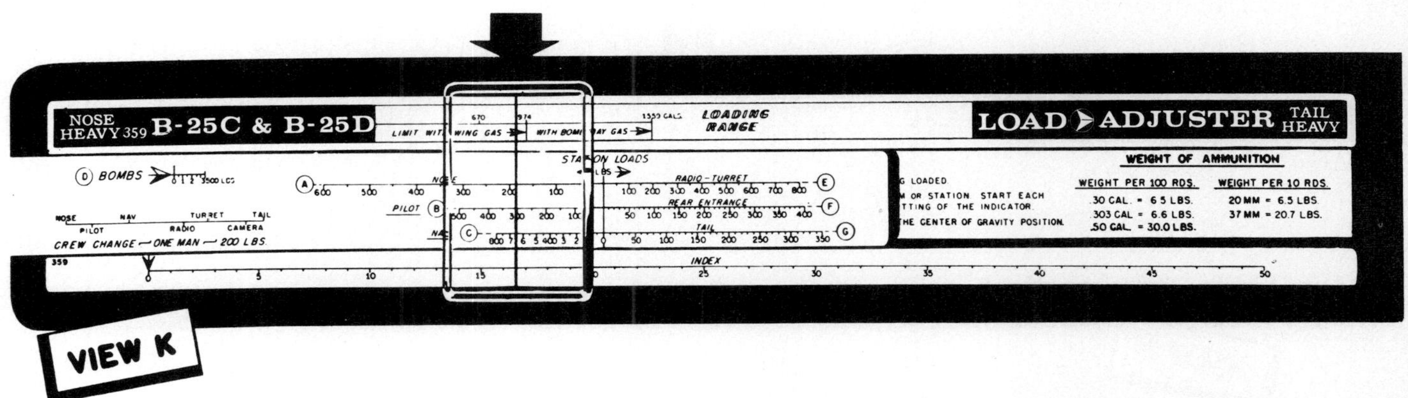

(8) Set slide to the compartment zero line as shown in view L.

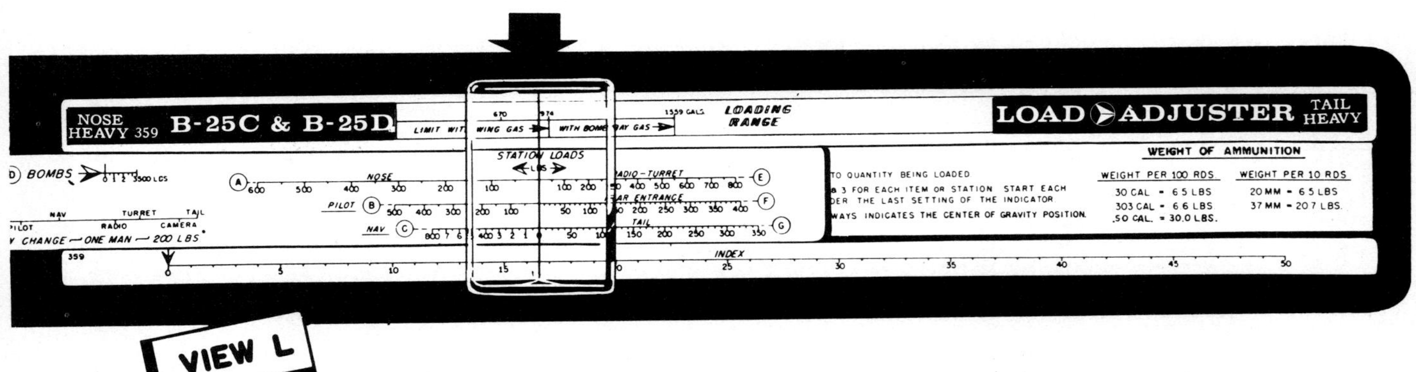

Move indicator until the hairline is over 800 on the "RADIO-TURRET" scale. This adds the balance moment of two men, special equipment and ammunition as loaded in this compartment and moves the index to 25.5 as shown in view M.

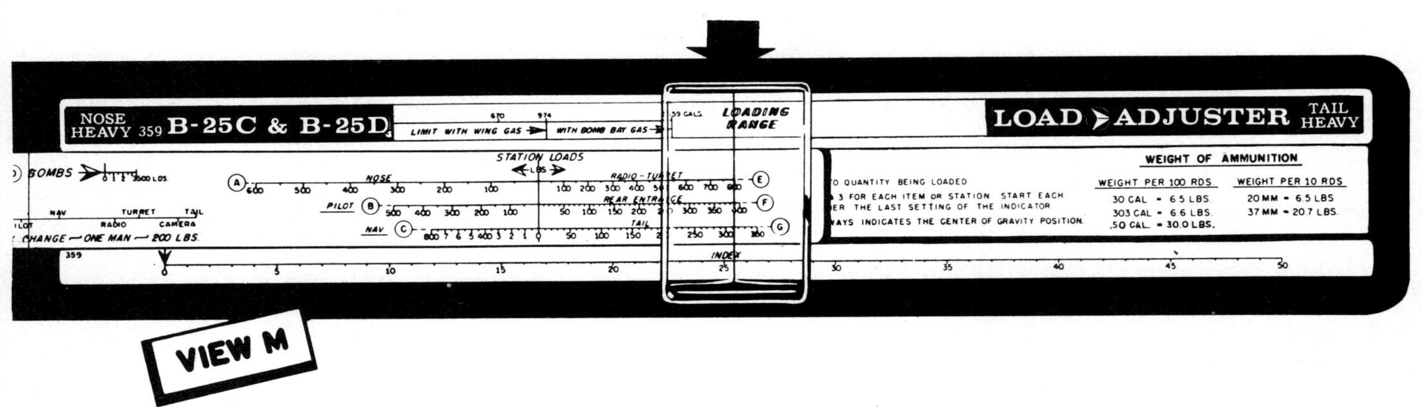

(9) Set slide to the compartment zero line as illustrated in view N.

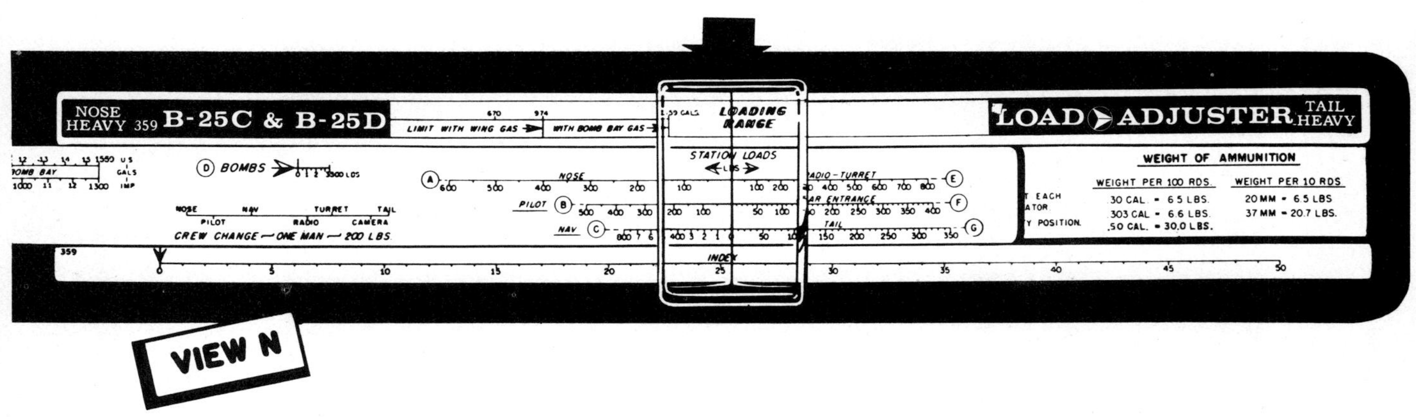

Move indicator until the hairline is over 400 on the "REAR ENTRANCE" scale. This adds the balance moment of photographic equipment and supplies, the photographer and brief cases as loaded in that compartment and moves the index to 34.5 as illustrated in view O.

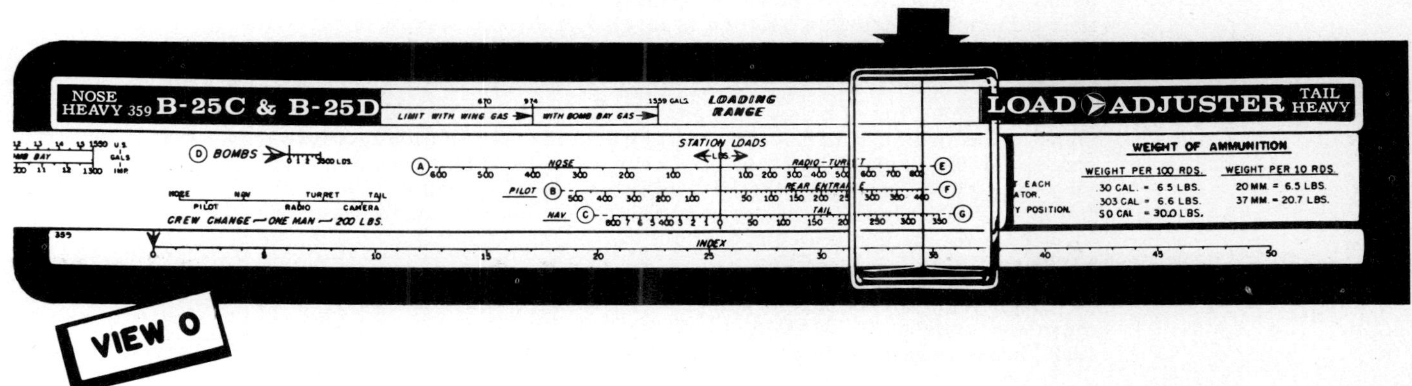

(10) Set slide to the compartment zero line as shown in view P.

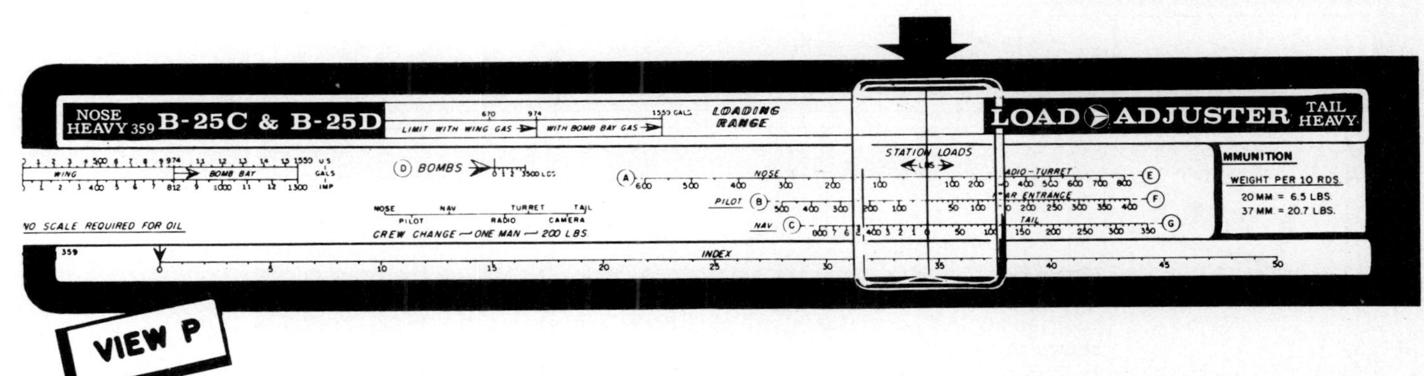

Move indicator until the hairline is over 125 on the "TAIL" compartment scale. This adds the balance moment of 125 pounds of equipment as loaded in the tail compartment and completes calculation of the balance moments of all items as initially loaded in the airplane. It has moved the airplane index to 38.0 as shown in view Q.

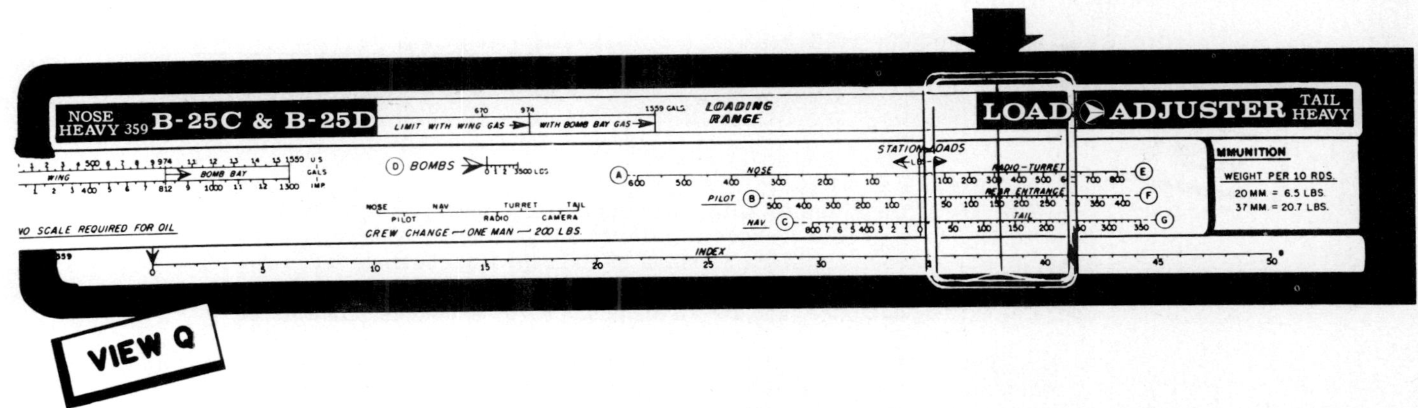

(11) Balance Correction.

(a) Adding the weights of all items loaded (paragraph 1.c. of this section) shows the gross load of the airplane well within allowable limits, and, as far as weight alone is concerned, the airplane may be flown. However, the load adjuster indicator hairline is located in the red portion of the loading range which ABSOLUTELY PROHIBITS any attempt to fly the airplane because of a dangerous tail-heavy condition.

(b) This "out of balance" condition may be corrected by shifting some of the load or a member of the crew from an aft position to a forward position in the airplane, the amount of change required being predetermined by a "tril shift" of load on the load adjuster. In this sample case, shifting the photographer from the camera position in the rear entrance compartment forward to the navigator's compartment will bring the airplane balance well within cg limits.

(c) With the indicator hairline remaining on the last index (38.0), move the slide until "CAMERA" of the "CREW CHANGE-ONE MAN-200 LBS" scale is under the indicator hairline as shown in view R.

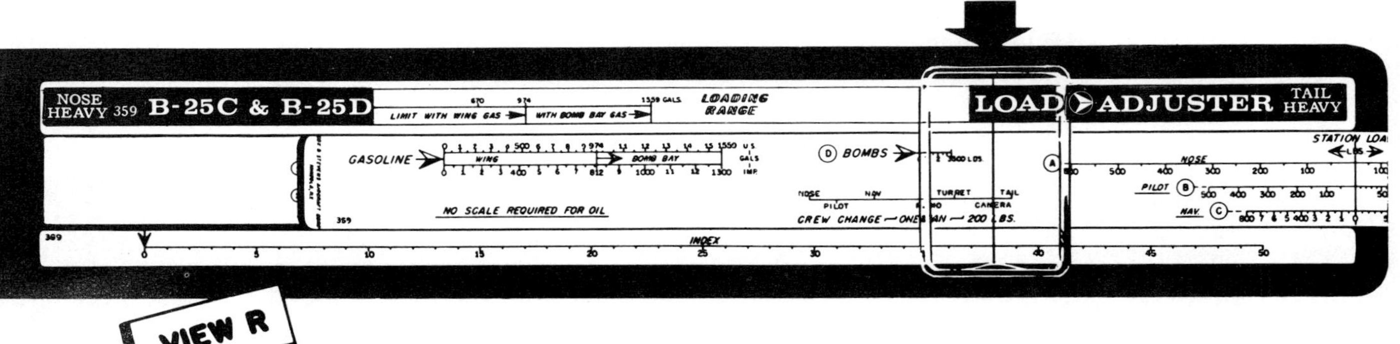

VIEW R

Move the indicator until its hairline is over "NAV." on the crew change scale. This changes the balance moment of one man (200 lb) from the camera position to the navigator's compartment, and moves the index to 32.7 as shown in view S.

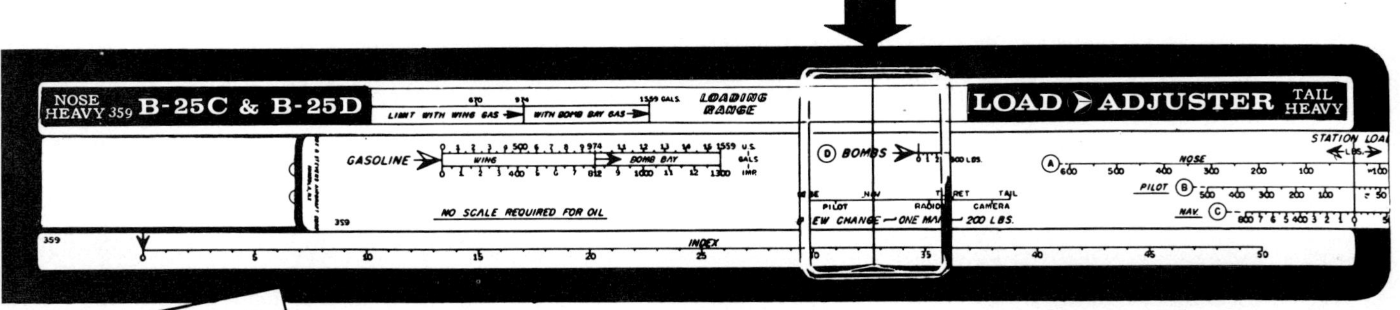

VIEW S

(12) The airplane is now within permissible cg limits, and is satisfactory for flight. It is entirely permissible to shift miscellaneous cargo, baggage, or other items within the airplane to make the final balance fall within approved limits. The load adjusting instrument is exceedingly simple to operate, equally accurate, and the airplane should be loaded as it indicates.

CAUTION: Do NOT shift or dispose of any load without first predetermining (by use of the load adjuster) that the balance will remain within limits after the change is made.

SECTION III

FLIGHT OPERATION DATA

1. <u>Determining Gross Weight.</u>

Secure gross weight from Form F, Weight & Balance Clearance.

2. <u>Flight Planning.</u>

The following outline may be used as a guide to assist personnel in the use of the FLIGHT OPERATION INSTRUCTION CHART for flight planning purposes.

<u>a.</u> If the flight plan calls for a continuous flight where the desired cruising power and air speed are reasonably constant after take-off and climb to 5000 feet, the fuel required and flight time may be computed as a "single section flight."

(1) Within the limits of the airplane, the fuel required and flying time for a given mission depend largely upon the speed desired. With all other factors remaining equal in an airplane, speed is obtained at a sacrifice of range, and range is obtained at a sacrifice of speed. The speed is usually determined after considering the urgency of the flight plotted against the range required. The time of take-off is adjusted so as to have the flight arrive at its destination at the predetermined time.

(2) Select the FLIGHT OPERATION INSTRUCTION CHART for the gross weight to be used at take-off. Locate the largest figure entered under "gph" (gallons per hour) in column I on the lower half of the chart. Multiply this figure by the number and/or fraction of hours desired for reserve fuel. Add the resulting figure to the number of gallons set forth in footnote No. 2, and subtract the total from the amount of fuel in the airplane prior to starting of engine. The figure obtained as a result of this computation will represent the amount of gasoline available and applicable for flight planning purposes on the RANGE IN AIR MILES section of the FLIGHT OPERATION INSTRUCTION CHART.

(3) Select a figure in the fuel column equal to, or the next entry less than, the available amount of fuel in the airplane, as determined in paragraph 2.<u>a.</u>(2) above. Move horizontally to the right or left and select a figure equal to, or the next entry greater than, the air miles (with no wind) to be flown. Operating values contained in the column number in which this figure appears, represent the highest cruising speed possible <u>at the range desired</u>; however, the airplane may be operated in accordance with values contained under OPERATING DATA in <u>any</u> column of a higher number, with the flight plan being completed at a sacrifice of speed but at an increase in fuel economy.

(4) Using the same column number selected by applications of instructions contained in **paragraph 2.<u>a.</u>**(3) above, determine the indicated air speed (in mph or knots, whichever is applicable to the calibration of instruments in the airplane) and gallons per hour listed at sea level in the lower section of the chart under the subtitle OPERATING DATA. Divide this IAS into the air miles to be flown and obtain the calculated flight duration in minutes, which can then be converted into hours and minutes and deducted from the desired arrival time at destination in order to obtain the take-off time (without consideration for wind). To allow for wind, use the above IAS as ground speed and calculate a new corrected ground speed with the aid of a flight calculator or by a navigator's triangle of velocities.

(5) The airplane and engine operating values listed below OPERATING DATA in any single numbered column are calculated to give constant miles per gallon at any altitude listed. Therefore, the airplane may be operated at any altitude and at the corresponding set of values given, so long as they are in the same column listing the range desired.

 <u>CAUTION:</u> Ranges listed in column I under "Max. Cont. Power" are correct only at the altitude given in footnote 1, and the engine and airplane operating data listed under OPERATING DATA will give constant miles per gallon if operation is consistent with values set opposite the listed altitudes.

(6) The flight plan may be readily changed at any time enroute, and the chart will show the balance of range at various cruising powers by following the INSTRUCTIONS FOR USING CHART printed on each page.

(7) Multiple charts are provided to give accurate data for operation at different gross weights, different external loads, and/or different combinations of engine use, such as single engine operation. Extreme caution should be exercised to assure selection of the correct chart applicable to the specific operating condition.

<u>b.</u> If the original flight plan calls for a mission requiring changes in power, speed, gross load, or external load, in accordance with the titles shown at the top of each chart provided, the total flight should be broken down into a series of individual short flights, each computed as outlined in paragraph 2.<u>b.</u> in its entirety, and then added together to make up the total flight and its requirements.

FLIGHT OPERATION INSTRUCTION CHART

MODEL (S)
B-25C
B-25D

SHEET......1......OF......8......SHEETS

EXTERNAL LOAD ITEMS
NONE

GR. WT...........34,000...........TO...........30,000...........POUNDS

CONDITION	R.P.M.	M.P. (IN HG.)	BLOWER POSITION	MIXTURE POSITION	DURATION IN. MIN.
TAKE-OFF	2600	44.0	LOW	FULL RICH	5
MILITARY POWER	2600	42 / 41	LOW / HIGH	FULL RICH	5
ENGINE (S)	R-2600-13				

INSTRUCTIONS FOR USING CHART: Select figure in fuel column equal to or less than total amount of fuel in airplane. Move horizontally to the right or left and select a figure equal to or greater than the air miles to be flown. Vertically below and opposite desired cruising altitude read optimum cruising conditions. *NOTES:* (A) Avoid continuous cruising in Column I except in emergency. (B) Columns (II, III, IV & V) toward the right progressively give increase in range at sacrifice in speed. (C) Manifold Pressure (M.P.), Gallons Per Hour (G.P.H.), and True Airspeed are approximate maximum values for reference. (D) For quick reference, take-off and military power data are listed in the upper left corner of chart.

ALTERNATE CRUISING CONDITIONS

(NO WIND) — (NO RESERVE FUEL ALLOWANCE)

Range in Air Miles

I (MAX. CONT. POWER) STATUTE	I NAUTICAL	FUEL U.S.GALS. [2]	II STATUTE	II NAUTICAL	III STATUTE	III NAUTICAL	IV STATUTE	IV NAUTICAL	FUEL IMP. GALS. [2]	V (MAX. RANGE) STATUTE	V NAUTICAL
1110	960	1520	1870	1630	2160	1880	2460	2140	1265	2750	2390
1020	890	1400	1730	1500	2000	1730	2260	1970	1165	2540	2200
950	820	1300	1600	1390	1850	1610	2100	1820	1082	2350	2040
870	760	1200	1480	1280	1710	1480	1940	1690	999	2170	1880
800	700	1100	1360	1180	1570	1360	1780	1540	916	1990	1730
730	630	1000	1230	1070	1420	1240	1620	1400	833	1810	1570
660	570	900	1110	960	1280	1110	1460	1260	750	1630	1410
580	510	800	990	860	1140	990	1290	1120	666	1450	1260
510	440	700	860	750	1000	870	1130	980	583	1270	1100
440	380	600	740	640	860	740	970	840	500	1090	940
360	320	500	620	540	710	620	810	700	417	900	790
290	250	400	490	430	570	490	650	560	333	720	630

OPERATING DATA

R.P.M.	M.P. IN.HG.	T.A.S. M.P.H.	T.A.S. KNOTS	U.S. G.P.H.	IMP. G.P.H.	DENSITY ALT. IN FEET	R.P.M.	M.P. IN.HG.	I.A.S. M.P.H.	I.A.S. KNOTS	U.S. G.P.H.	IMP. G.P.H.	R.P.M.	M.P. IN.HG.	I.A.S. M.P.H.	I.A.S. KNOTS	U.S. G.P.H.	IMP. G.P.H.	R.P.M.	M.P. IN.HG.	I.A.S. M.P.H.	I.A.S. KNOTS	U.S. G.P.H.	IMP. G.P.H.	DENSITY ALT. IN FEET	R.P.M.	M.P. IN.HG.	I.A.S. M.P.H.	I.A.S. KNOTS	U.S. G.P.H.	IMP. G.P.H.
						30000																			30000						
						25000																			25000						
2400	F.T.	260	225	300	250	20000	2200	28	170	147	215	179	2100	27.5	160	138	155	129							20000						
2400	F.T.	280	242	375	312	15000	2100	32.5	195	169	230	191	2250	25.5	190	165	165	137	2050	23.5	175	152			15000						
2400	39	275	238	365	304	12000	2200	29	210	182	220	183	2100	27.5	200	173	175	146	2000	26	185	160			12000						
2400	F.T.	275	238	330	275	9000	2100	30.5	215	186	215	179	2050	28.5	205	178	175	146	2000	27.5	195	169			9000	1850	26	170	147	110	92
2400	38	275	238	375	312	6000	2050	30.5	215	186	210	175	2000	29	210	182	170	142	2000	28	200	173			6000	1800	27	175	152	110	92
2400	38	265	229	355	295	3000	2100	31.5	220	190	210	175	2000	29.5	210	182	160	133	1950	28.5	200	173	135	112	3000	1700	28	180	156	105	87
2400	38	255	221	340	283	S.L.	2050	30.5	220	190	175	146	2000	30	210	182	150	125	1950	29.0	200	173	130	108	S.L.	1650	29	180	156	105	87

LEGEND

1 Range values at __14,000__ ft. only

2 Allow __19__ U. S. gals., __16__ Imp. gals. for warm up, take-off to __5,000__ feet altitude and climb. Use fuel from tanks in the following order_____________

Return fuel flows to tank

REFER TO "SPECIFIC ENGINE FLIGHT CHART" FOR ADDITIONAL ENGINE OPERATION DATA.

BOLD NUMBERS: Use Auto-Rich
LIGHT ITALICS: Use Auto-Lean
WITH TWO SPEED BLOWER: Use high blower above heavy line only

I.A.S.: Indicated Air Speed
T.A.S.: True Air Speed
M.P.: Manifold Pressure (In. Hg.)
U.S.G.P.H.: U. S. Gallons Per Hour
IMP.G.P.H.: Imperial Gallons Per Hour
F.T.: Full Throttle

UNDERLINED FIGURES ARE PRELIMINARY: SUBJECT TO REVISION AFTER FLIGHT CHECK

MODEL (S) — B-25C / B-25D

FLIGHT OPERATION INSTRUCTION CHART

SHEET....2....OF....8....SHEETS

EXTERNAL LOAD ITEMS

NONE

GR. WT......30,000......TO......26,000......POUNDS

CONDITION	R.P.M.	M.P. (IN HG.)	BLOWER POSITION	MIXTURE POSITION	DURATION IN. MIN.
TAKE-OFF	2600	44.0	LOW	FULL RICH	5
MILITARY POWER	2600	42 / 41	LOW / HIGH	FULL RICH	5
ENGINE (S)	R-2600-13				

INSTRUCTIONS FOR USING CHART: Select figure in fuel column equal to or less than total amount of fuel in airplane. Move horizontally to the right or left and select a figure equal to or greater than the air miles to be flown. Vertically below and opposite desired cruising altitude read optimum cruising conditions. **NOTES:** (A) Avoid continuous cruising in Column I except in emergency. (B) Columns (II, III, IV & V) toward the right progressively give increase in range at sacrifice in speed. (C) Manifold Pressure (M.P.), Gallons Per Hour (G.P.H.), and True Airspeed are approximate maximum values for reference. (D) For quick reference, take-off and military power data are listed in the upper left corner of chart.

ALTERNATE CRUISING CONDITIONS

(NO WIND) (NO RESERVE FUEL ALLOWANCE)

RANGE IN AIR MILES

I (MAX. CONT. POWER) STATUTE	I NAUTICAL	FUEL U.S. GALS. 2	II STATUTE	II NAUTICAL	III STATUTE	III NAUTICAL	IV STATUTE	IV NAUTICAL	FUEL IMP. GALS. 2	V (MAX. RANGE) STATUTE	V NAUTICAL
760	660	940	1280	1110	1480	1280	1680	1460	783	1880	1630
640	550	800	1090	940	1260	1090	1430	1240	666	1600	1390
560	480	700	950	820	1100	950	1250	1080	583	1400	1210
480	410	600	810	700	940	810	1070	920	500	1200	1040
400	340	500	680	590	780	670	890	770	417	1000	860
320	270	400	540	470	630	540	710	610	333	800	690
240	200	300	400	340	470	400	530	460	250	600	520
160	130	200	270	230	310	270	350	300	167	400	340
80	60	100	130	110	150	130	170	140	83	200	170

OPERATING DATA — Column I (MAX. CONT. POWER)

DENSITY ALT. IN FEET	R.P.M.	M.P. IN.HG.	T.A.S. M.P.H.	T.A.S. KNOTS	U.S.G.P.H.	IMP.G.P.H.
30000						
25000						
20000	2400	F.T.	275	238	300	250
15000	2400	F.T.	285	247	395	328
12000	2400	39	280	242	365	304
9000	2400	F.T.	280	242	330	275
6000	2400	38	280	242	370	308
3000	2400	38	270	234	355	295
S.L.	2400	38	260	225	340	283

OPERATING DATA — Column II

DENSITY ALT. IN FEET	R.P.M.	M.P. IN.HG.	I.A.S. M.P.H.	I.A.S. KNOTS	U.S.G.P.H.	IMP.G.P.H.
20000	2150	27.5	180	156	205	171
15000	2050	31.5	200	173	215	179
12000	2100	28	210	182	195	162
9000	2000	28.5	210	182	160	133
6000	2000	29	215	186	170	142
3000	2000	30	220	190	175	146
S.L.	2000	29.5	215	186	140	117

OPERATING DATA — Column III

DENSITY ALT. IN FEET	R.P.M.	M.P. IN.HG.	I.A.S. M.P.H.	I.A.S. KNOTS	U.S.G.P.H.	IMP.G.P.H.
20000	2100	27	175	152	150	125
15000	2150	25	195	169	150	125
12000	2050	27.5	205	178	170	142
9000	2000	27.5	205	178	145	121
6000	2000	28.5	205	178	150	125
3000	2000	29	210	182	145	121
S.L.	1900	29	205	177	130	108

OPERATING DATA — Column IV

DENSITY ALT. IN FEET	R.P.M.	M.P. IN.HG.	I.A.S. M.P.H.	I.A.S. KNOTS	U.S.G.P.H.	IMP.G.P.H.
15000	1850	22	160	138	115	96
12000	1900	25	180	156	130	108
9000	1950	27	195	169	130	108
6000	1900	27	195	169	125	104
3000	1850	28	195	169	120	100
S.L.	1700	29	195	169	115	96

OPERATING DATA — Column V (MAX. RANGE)

DENSITY ALT. IN FEET	R.P.M.	M.P. IN.HG.	I.A.S. M.P.H.	I.A.S. KNOTS	U.S.G.P.H.	IMP.G.P.H.
9000	1750	25.5	170	147	100	83
6000	1650	26.5	165	143	100	83
3000	1450	27.5	160	138	85	71
S.L.	1400	29	165	143	85	71

LEGEND

1 Range values at __14,000__ ft. only

2 Allow __34__ U. S. gals., __28__ Imp. gals. for warm up, take-off to __5,000__ feet altitude and climb. Use fuel from tanks in the following order________________ Return fuel flows to tank

BOLD NUMBERS: Use Auto-Rich
LIGHT ITALICS: Use Auto-Lean
WITH TWO SPEED BLOWER: Use high blower above heavy line only

I.A.S.: Indicated Air Speed
T.A.S.: True Air Speed
M.P.: Manifold Pressure (In. Hg.)
U.S.G.P.H.: U. S. Gallons Per Hour
IMP.G.P.H.: Imperial Gallons Per Hour
F.T.: Full Throttle

REFER TO "SPECIFIC ENGINE FLIGHT CHART" FOR ADDITIONAL ENGINE OPERATION DATA.

UNDERLINED FIGURES ARE PRELIMINARY: SUBJECT TO REVISION AFTER FLIGHT CHECK

MODEL (S)
B-25 C
B-25 D

FLIGHT OPERATION INSTRUCTION CHART

SHEET....3....OF....8....SHEETS

EXTERNAL LOAD ITEMS
1 TORPEDO AND 8 WING BOMBS

GR. WT.......36,000.......TO.......32,000.........POUNDS

CONDITION	R.P.M.	M.P. (IN HG.)	BLOWER POSITION	MIXTURE POSITION	DURATION IN. MIN.
TAKE-OFF	2600	44.0	LOW	FULL RICH	5
MILITARY POWER	2600	42 / 41	LOW / HIGH	FULL RICH	5
ENGINE (S)	R-2600-13				

INSTRUCTIONS FOR USING CHART: Select figure in fuel column equal to or less than total amount of fuel in airplane. Move horizontally to the right or left and select a figure equal to or greater than the air miles to be flown. Vertically below and opposite desired cruising altitude read optimum cruising conditions. *NOTES:* (A) Avoid continuous cruising in Column I except in emergency. (B) Columns (II, III, IV & V) toward the right progressively give increase in range at sacrifice in speed. (C) Manifold Pressure (M.P.), Gallons Per Hour (G.P.H.), and True Airspeed are approximate maximum values for reference. (D) For quick reference, take-off and military power data are listed in the upper left corner of chart.

ALTERNATE CRUISING CONDITIONS

(NO WIND) (NO RESERVE FUEL ALLOWANCE)

I (MAX. CONT. POWER)		FUEL U.S.GALS. 2	II		III		IV		FUEL IMP. GALS. 2	V (MAX. RANGE)	
RANGE IN AIR MILES			RANGE IN AIR MILES		RANGE IN AIR MILES		RANGE IN AIR MILES			RANGE IN AIR MILES	
STATUTE	NAUTICAL		STATUTE	NAUTICAL	STATUTE	NAUTICAL	STATUTE	NAUTICAL		STATUTE	NAUTICAL
850	730	1380	1190	1030	1390	1200	1580	1370	1149	1770	1530
800	690	1300	1120	970	1300	1130	1480	1280	1082	1670	1450
740	640	1200	1040	900	1200	1040	1370	1190	999	1540	1330
680	590	1100	950	820	1100	950	1260	1090	916	1410	1220
620	530	1000	860	740	1000	860	1140	990	833	1280	1110
550	470	900	780	670	900	780	1030	890	750	1150	990
490	420	800	690	600	800	690	910	790	666	1020	880
430	370		600	520	700	600	800	690		900	780

OPERATING DATA — I

R.P.M.	M.P. IN.HG.	T.A.S. M.P.H.	T.A.S. KNOTS	U.S. G.P.H.	IMP. G.P.H.	DENSITY ALT. IN FEET
						30000
						25000
						20000
2400	F.T.	231	200	395	328	15000
2400	39	227	197	365	304	12000
2400	F.T.	234	203	330	275	9000
2400	38	237	206	370	308	6000
2400	38	230	199	355	295	3000
2400	38	219	190	340	283	S.L.

OPERATING DATA — II

R.P.M.	M.P. IN.HG.	I.A.S. M.P.H.	I.A.S. KNOTS	U.S. G.P.H.	IMP. G.P.H.
2100	32	150	130	235	195
2200	36.5	175	152	290	241
2200	31.5	185	160	270	225
2150	32	190	165	260	216
2150	33	195	169	260	216
2150	34	195	169	255	212

OPERATING DATA — III

R.P.M.	M.P. IN.HG.	I.A.S. M.P.H.	I.A.S. KNOTS	U.S. G.P.H.	IMP. G.P.H.
2200	29	170	147	220	183
2100	30	175	152	215	179
2100	30.5	180	156	215	179
2100	31.5	185	160	215	179
2050	32	190	165	205	171

OPERATING DATA — IV

R.P.M.	M.P. IN.HG.	I.A.S. M.P.H.	I.A.S. KNOTS	U.S. G.P.H.	IMP. G.P.H.	DENSITY ALT. IN FEET
						30000
						25000
						20000
						15000
2100	28	160	139	190	158	12000
2050	29	165	143	190	158	9000
2050	30	175	152	185	154	6000
2050	30.5	180	165	180	150	3000
2050	31	180	165	175	146	S.L.

OPERATING DATA — V

R.P.M.	M.P. IN.HG.	I.A.S. M.P.H.	I.A.S. KNOTS	U.S. G.P.H.	IMP. G.P.H.
2000	28	150	130	150	125
2000	28	155	134	145	121
2000	29	165	143	145	121
2000	29.5	170	147	145	121

LEGEND

1 Range values at ____14,000____ ft. only

2 Allow ____59____ U. S. gals., ____49____ Imp. gals. for warm up, take-off to ____5,000____ feet altitude and climb. Use fuel from tanks in the following order ____________ Return fuel flows to tank

REFER TO "SPECIFIC ENGINE FLIGHT CHART" FOR ADDITIONAL ENGINE OPERATION DATA.

BOLD NUMBERS: Use Auto-Rich
LIGHT ITALICS: Use Auto-Lean
WITH TWO SPEED BLOWER: Use high blower above heavy line only

I.A.S.: Indicated Air Speed
T.A.S.: True Air Speed
M.P.: Manifold Pressure (In. Hg.)
U.S.G.P.H.: U. S. Gallons Per Hour
IMP.G.P.H.: Imperial Gallons Per Hour
F.T.: Full Throttle

FLIGHT OPERATION INSTRUCTION CHART

MODEL (S)
B-25 C
B-25 D

SHEET 4 OF 8 SHEETS

EXTERNAL LOAD ITEMS
1 TORPEDO AND 8 WING BOMBS

GR. WT. 32,000 TO 28,000 POUNDS

CONDITION	R.P.M.	M.P. (IN HG.)	BLOWER POSITION	MIXTURE POSITION	DURATION IN. MIN.
TAKE-OFF	2600	44.0	LOW	FULL RICH	5
MILITARY POWER	2600	42 / 41	LOW / HIGH	FULL RICH	5
ENGINE (S)	R-2600-13				

INSTRUCTIONS FOR USING CHART: Select figure in fuel column equal to or less than total amount of fuel in airplane. Move horizontally to the right or left and select a figure equal to or greater than the air miles to be flown. Vertically below and opposite desired cruising altitude read optimum cruising conditions. NOTES: (A) Avoid continuous cruising in Column I except in emergency. (B) Columns (II, III, IV & V) toward the right progressively give increase in range at sacrifice in speed. (C) Manifold Pressure (M.P.), Gallons Per Hour (G.P.H.), and True Airspeed are approximate maximum values for reference. (D) For quick reference, take-off and military power data are listed in the upper left corner of chart.

ALTERNATE CRUISING CONDITIONS

(NO WIND)   (NO RESERVE FUEL ALLOWANCE)

I (MAX. CONT. POWER)		FUEL U.S.GALS. 2	II		III		IV		FUEL IMP. GALS. 2	V (MAX. RANGE)	
RANGE IN AIR MILES			RANGE IN AIR MILES		RANGE IN AIR MILES		RANGE IN AIR MILES			RANGE IN AIR MILES	
STATUTE	NAUTICAL		STATUTE	NAUTICAL	STATUTE	NAUTICAL	STATUTE	NAUTICAL		STATUTE	NAUTICAL
550	470	925	900	780	1040	900	1180	1020	771	1320	1140
470	400	800	780	670	900	780	1020	880	666	1140	990
410	350	700	680	590	790	680	890	770	583	1000	870
350	300	600	580	500	680	590	770	660	500	860	740
290	250	500	490	420	560	480	640	550	417	710	610
230	200	400	390	330	450	390	510	440	333	570	490
170	140	300	290	250	340	290	380	330	250	430	370
110	95	200	190	160	220	190	250	210	167	280	240
60	50	100	100	80	110	90	120	100	83	140	120

OPERATING DATA						DENSITY ALT. IN FEET	OPERATING DATA						OPERATING DATA						OPERATING DATA						DENSITY ALT. IN FEET	OPERATING DATA					
R.P.M.	M.P. IN.HG.	T.A.S. M.P.H.	T.A.S. KNOTS	U.S. G.P.H.	IMP. G.P.H.		R.P.M.	M.P. IN.HG.	I.A.S. M.P.H.	I.A.S. KNOTS	U.S. G.P.H.	IMP. G.P.H.	R.P.M.	M.P. IN.HG.	I.A.S. M.P.H.	I.A.S. KNOTS	U.S. G.P.H.	IMP. G.P.H.	R.P.M.	M.P. IN.HG.	I.A.S. M.P.H.	I.A.S. KNOTS	U.S. G.P.H.	IMP. G.P.H.		R.P.M.	M.P. IN.HG.	I.A.S. M.P.H.	I.A.S. KNOTS	U.S. G.P.H.	IMP. G.P.H.
						30000																			30000						
						25000																			25000						
						20000																			20000						
2400	F.T.	240	208	395	328	15000	2150	33.5	170	147	250	208	2100	32.5	170	147	235	195	2050	31.5	160	139	215	179	15000						
2400	39	235	204	365	304	12000	2200	29	180	156	225	187	2100	28	175	152	195	162	2050	27	170	147	165	137	12000						
2400	F.T.	240	208	330	275	9000	2100	30.5	190	165	230	191	2050	29	180	156	190	158	2000	28.5	175	152	160	133	9000	1950	29	155	134	130	108
2400	38	245	212	370	308	6000	2100	31	190	165	220	183	2050	30	185	160	190	158	2000	28.5	175	152	155	129	6000	1900	29	160	139	125	104
2400	38	235	204	355	295	3000	2100	31.5	195	169	215	179	2050	30.5	185	160	175	146	2000	29	180	156	150	125	3000	1800	29	160	139	120	100
2400	38	225	195	240	200	S.L.	2050	32.5	200	173	210	175	2000	31	190	165	165	137	2000	29.5	180	156	140	117	S.L.	1800	29	165	143	120	100

LEGEND

1 Range values at 14,000 ft. only
2 Allow 49 U. S. gals., 41 Imp. gals. for warm up, take-off to 5,000 feet altitude and climb. Use fuel from tanks in the following order __________
Return fuel flows to tank

BOLD NUMBERS: Use Auto-Rich
LIGHT ITALICS: Use Auto-Lean
WITH TWO SPEED BLOWER: Use high blower above heavy line only

I.A.S.: Indicated Air Speed
T.A.S.: True Air Speed
M.P.: Manifold Pressure (In. Hg.)
U.S.G.P.H.: U. S. Gallons Per Hour
IMP.G.P.H.: Imperial Gallons Per Hour
F.T.: Full Throttle

REFER TO "SPECIFIC ENGINE FLIGHT CHART" FOR ADDITIONAL ENGINE OPERATION DATA.

UNDERLINED FIGURES ARE PRELIMINARY: SUBJECT TO REVISION AFTER FLIGHT CHECK

FLIGHT OPERATION INSTRUCTION CHART

MODEL (S) B-25 C / B-25 D

SHEET 5 OF 8 SHEETS

EXTERNAL LOAD ITEMS
1 TORPEDO OR 8 WING BOMBS

GR. WT. 36,000 TO 31,000 POUNDS

CONDITION	R.P.M.	M.P. (IN HG.)	BLOWER POSITION	MIXTURE POSITION	DURATION IN. MIN.
TAKE-OFF	2600	44.0	LOW	FULL RICH	5
MILITARY POWER	2600	42 / 41	LOW / HIGH	FULL RICH	5
ENGINE (S)	R-2600-13				

INSTRUCTIONS FOR USING CHART: Select figure in fuel column equal to or less than total amount of fuel in airplane. Move horizontally to the right or left and select a figure equal to or greater than the air miles to be flown. Vertically below and opposite desired cruising altitude read optimum cruising conditions. **NOTES:** (A) Avoid continuous cruising in Column I except in emergency. (B) Columns (II, III, IV & V) toward the right progressively give increase in range at sacrifice in speed. (C) Manifold Pressure (M.P.), Gallons Per Hour (G.P.H.), and True Airspeed are approximate maximum values for reference. (D) For quick reference, take-off and military power data are listed in the upper left corner of chart.

ALTERNATE CRUISING CONDITIONS

(NO WIND) (NO RESERVE FUEL ALLOWANCE)

I (MAX. CONT. POWER)		FUEL U.S.GALS. 2	II		III		IV		FUEL IMP. GALS. *2	V (MAX. RANGE)	
RANGE IN AIR MILES			RANGE IN AIR MILES		RANGE IN AIR MILES		RANGE IN AIR MILES			RANGE IN AIR MILES	
STATUTE	NAUTICAL		STATUTE	NAUTICAL	STATUTE	NAUTICAL	STATUTE	NAUTICAL		STATUTE	NAUTICAL
1100	950	1520	1570	1360	1810	1570	2060	1780	1265	2300	1990
1010	880	1400	1450	1260	1670	1450	1890	1640	1165	2120	1840
940	820	1300	1350	1170	1550	1350	1760	1530	1082	1960	1710
870	750	1200	1240	1080	1430	1240	1620	1410	999	1810	1570
800	690	1100	1140	990	1310	1140	1490	1290	916	1660	1440
720	630	1000	1040	900	1190	1040	1350	1170	833	1510	1310
650	570	900	930	810	1070	930	1220	1060	750	1360	1180

OPERATING DATA

DENSITY ALT. IN FEET	I — R.P.M.	M.P. IN.HG.	T.A.S. M.P.H.	T.A.S. KNOTS	U.S. G.P.H.	IMP. G.P.H.	II — R.P.M.	M.P. IN.HG.	I.A.S. M.P.H.	I.A.S. KNOTS	U.S. G.P.H.	IMP. G.P.H.	III — R.P.M.	M.P. IN.HG.	I.A.S. M.P.H.	I.A.S. KNOTS	U.S. G.P.H.	IMP. G.P.H.	IV — R.P.M.	M.P. IN.HG.	I.A.S. M.P.H.	I.A.S. KNOTS	U.S. G.P.H.	IMP. G.P.H.	V — R.P.M.	M.P. IN.HG.	I.A.S. M.P.H.	I.A.S. KNOTS	U.S. G.P.H.	IMP. G.P.H.
30000																														
25000																														
20000																														
15000	2400	F.T.	248	215	375	241	2100	33.5	165	143	240	200	2100	33	165	143	230	191	2050	32	160	139	210	175						
12000	2400	39	245	212	365	230	2200	36.5	185	160	290	241	2150	28.5	175	152	200	167	2050	27.5	170	147	170	142						
9000	2400	F.T.	252	220	330	258	2150	31	190	165	245	204	2050	29.5	180	156	200	167	2000	28.5	175	152	170	142	1950	27	155	134	135	112
6000	2400	38	254	222	370	250	2100	31.5	195	169	240	200	2050	30	185	160	200	167	2000	29	180	156	170	142	1950	27.5	165	143	135	112
3000	2400	38	244	211	355	241	2100	32	200	173	230	191	2050	31	190	165	190	158	2000	30	185	160	165	137	1950	28.5	170	147	130	108
S.L.	2400	38	233	202	340	229	2100	32.5	200	173	220	183	2050	31.5	195	169	180	150	2000	30.5	185	160	155	129	1900	29	175	152	130	108

LEGEND

1 Range values at 14,000 ft. only

2 Allow 39 U. S. gals., 32 Imp. gals. for warm up, take-off to 5,000 feet altitude and climb. Use fuel from tanks in the following order _______ Return fuel flows to tank

BOLD NUMBERS: Use Auto-Rich
LIGHT ITALICS: Use Auto-Lean
WITH TWO SPEED BLOWER: Use high blower above heavy line only

I.A.S.: Indicated Air Speed
T.A.S.: True Air Speed
M.P.: Manifold Pressure (In. Hg.)
U.S.G.P.H.: U. S. Gallons Per Hour
IMP.G.P.H.: Imperial Gallons Per Hour
F.T.: Full Throttle

REFER TO "SPECIFIC ENGINE FLIGHT CHART" FOR ADDITIONAL ENGINE OPERATION DATA.

UNDERLINED FIGURES ARE PRELIMINARY: SUBJECT TO REVISION AFTER FLIGHT CHECK

FLIGHT OPERATION INSTRUCTION CHART

MODEL (S)
B-25 C
B-25 D

SHEET......6......OF......8......SHEETS

EXTERNAL LOAD ITEMS
1 TORPEDO OR 8 WING BOMBS

GR. WT.........31,000.........TO.........26,000.........POUNDS

CONDITION	R.P.M.	M.P. (IN HG.)	BLOWER POSITION	MIXTURE POSITION	DURATION IN. MIN.
TAKE-OFF	2600	44.0	LOW	FULL RICH	5
MILITARY POWER	2600	42 / 41	LOW / HIGH	FULL RICH	5

ENGINE (S)   R-2600-13

INSTRUCTIONS FOR USING CHART: Select figure in fuel column equal to or less than total amount of fuel in airplane. Move horizontally to the right or left and select a figure equal to or greater than the air miles to be flown. Vertically below and opposite desired cruising altitude read optimum cruising conditions. **NOTES:** (A) Avoid continuous cruising in Column I except in emergency. (B) Columns (II, III, IV & V) toward the right progressively give increase in range at sacrifice in speed. (C) Manifold Pressure (M.P.), Gallons Per Hour (G.P.H.), and True Airspeed are approximate maximum values for reference. (D) For quick reference, take-off and military power data are listed in the upper left corner of chart.

ALTERNATE CRUISING CONDITIONS

(NO WIND)   (NO RESERVE FUEL ALLOWANCE)

I (MAX. CONT. POWER) RANGE IN AIR MILES		FUEL U.S.GALS. 2	II RANGE IN AIR MILES		III RANGE IN AIR MILES		IV RANGE IN AIR MILES		FUEL IMP. GALS. 2	V (MAX. RANGE) RANGE IN AIR MILES	
STATUTE	NAUTICAL		STATUTE	NAUTICAL	STATUTE	NAUTICAL	STATUTE	NAUTICAL		STATUTE	NAUTICAL
620	530	940	1110	960	1270	1100	1430	1240	783	1600	1330
520	450	800	950	820	1080	940	1220	1060	666	1360	1180
460	400	700	830	720	950	820	1070	930	583	1190	1030
390	340	600	710	620	810	700	910	790	500	1020	880
330	280	500	590	510	680	590	760	660	417	850	740
260	230	400	470	410	540	470	610	530	333	680	590
200	170	300	350	310	410	350	460	400	250	510	440
130	110	200	240	200	270	230	300	260	167	340	300
60	50	100	110	100	130	110	150	130	83	170	140

OPERATING DATA — Column I

R.P.M.	M.P. IN.HG.	T.A.S. M.P.H.	T.A.S. KNOTS	U.S. G.P.H.	IMP. G.P.H.	DENSITY ALT. IN FEET
						30000
						25000
2400	F.T.	250	216	300	250	20000
2400	F.T.	260	225	375	312	15000
2400	39	254	220	365	304	12000
2400	F.T.	257	222	330	275	9000
2400	38	259	224	370	308	6000
2400	38	248	215	355	295	3000
2400	38	239	207	340	283	S. L.

OPERATING DATA — Column II

R.P.M.	M.P. IN.HG.	I.A.S. M.P.H.	I.A.S. KNOTS	U.S. G.P.H.	IMP. G.P.H.
2100	27	150	130	160	133
2200	25.5	180	156	170	142
2150	28.5	195	169	200	167
2050	29.5	195	169	190	158
2050	30	200	173	190	158
2050	30.5	200	173	185	154
2050	31.5	205	178	180	150

OPERATING DATA — Column III

R.P.M.	M.P. IN.HG.	I.A.S. M.P.H.	I.A.S. KNOTS	U.S. G.P.H.	IMP. G.P.H.
2000	26.5	145	126	145	121
2150	25	175	152	150	125
2100	27.5	185	160	175	146
2000	28.5	190	165	165	137
2000	29	190	165	165	137
2000	29.5	195	169	155	129
2000	30	200	173	155	129

OPERATING DATA — Column IV

R.P.M.	M.P. IN.HG.	I.A.S. M.P.H.	I.A.S. KNOTS	U.S. G.P.H.	IMP. G.P.H.	DENSITY ALT. IN FEET
						30000
						25000
						20000
2000	24	160	139	140	117	15000
2000	26.5	175	152	140	117	12000
2000	27	180	156	140	117	9000
1950	27.5	185	160	135	112.2	6000
1950	28.5	185	160	130	108	3000
1900	29	185	160	130	108	S. L.

OPERATING DATA — Column V

ALT. IN FEET	R.P.M.	M.P. IN.HG.	I.A.S. M.P.H.	I.A.S. KNOTS	U.S. G.P.H.	IMP. G.P.H.
30000						
25000						
20000						
15000						
12000						
9000	1800	26	155	130	105	87
6000	1750	27	160	139	105	87
3000	1600	28	160	139	100	83
S. L.	1500	29	160	139	95	79

LEGEND

1. Range values at 14,000 ft. only
2. Allow 34 U. S. gals., 28 Imp. gals. for warm up, take-off to 5,000 feet altitude and climb. Use fuel from tanks in the following order __________
Return fuel flows to tank

REFER TO "SPECIFIC ENGINE FLIGHT CHART" FOR ADDITIONAL ENGINE OPERATION DATA.

BOLD NUMBERS: Use Auto-Rich
LIGHT ITALICS: Use Auto-Lean
WITH TWO SPEED BLOWER: Use high blower above heavy line only

I.A.S.: Indicated Air Speed
T.A.S.: True Air Speed
M.P.: Manifold Pressure (In. Hg.)
U.S.G.P.H.: U. S. Gallons Per Hour
IMP.G.P.H.: Imperial Gallons Per Hour
F.T.: Full Throttle

UNDERLINED FIGURES ARE PRELIMINARY: SUBJECT TO REVISION AFTER FLIGHT CHECK

FLIGHT OPERATION INSTRUCTION CHART
(SINGLE ENGINE OPERATION)

MODEL (S)
B-25 C
B-25 D

GR. WT......28,000......TO......26,000......POUNDS

SHEET.....7.....OF......8....SHEETS

EXTERNAL LOAD ITEMS
NONE

CONDITION	R.P.M.	M.P. (IN HG.)	BLOWER POSITION	MIXTURE POSITION	DURATION IN. MIN.
TAKE-OFF	2600	44.0	LOW	FULL RICH	5
MILITARY POWER	2600	42 / 41	LOW / HIGH	FULL RICH	5
ENGINE (S)	R-2600-13 (1)				

INSTRUCTIONS FOR USING CHART: Select figure in fuel column equal to or less than total amount of fuel in airplane. Move horizontally to the right or left and select a figure equal to or greater than the air miles to be flown. Vertically below and opposite desired cruising altitude read optimum cruising conditions. *NOTES:* (A) Avoid continuous cruising in Column I except in emer-gency. (B) Columns (II, III, IV & V) toward the right progressively give increase in range at sacrifice in speed. (C) Manifold Pressure (M.P.), Gallons Per Hour (G.P.H.), and True Airspeed are approximate maximum values for reference. (D) For quick reference, take-off and military power data are listed in the upper left corner of chart.

ALTERNATE CRUISING CONDITIONS

(NO WIND)　　　　(NO RESERVE FUEL ALLOWANCE)

I (MAX. CONT. POWER) RANGE IN AIR MILES		FUEL U.S.GALS. 2	II RANGE IN AIR MILES		III RANGE IN AIR MILES		IV RANGE IN AIR MILES		FUEL IMP. GALS. 2	V (MAX. RANGE) RANGE IN AIR MILES	
STATUTE	NAUTICAL		STATUTE	NAUTICAL	STATUTE	NAUTICAL	STATUTE	NAUTICAL		STATUTE	NAUTICAL
940	810	900			1040	900	1170	1010	750	1300	1130
830	720	800			920	800	1040	900	666	1160	1000
730	630	700			810	700	910	790	583	1010	880
620	540	600			690	600	780	670	500	870	750
520	450	500			580	500	650	560	417	720	630
410	360	400			460	400	520	450	333	590	510
310	270	300			340	300	390	340	250	430	370
210	180	200			230	200	260	220	167	290	250
100	90	100			120	100	130	110	83	140	120

OPERATING DATA						DENSITY ALT. IN FEET	OPERATING DATA						OPERATING DATA						OPERATING DATA						DENSITY ALT. IN FEET	OPERATING DATA					
R.P.M.	M.P. IN.HG.	T.A.S. M.P.H.	T.A.S. KNOTS	U.S. G.P.H.	IMP. G.P.H.		R.P.M.	M.P. IN.HG.	I.A.S. M.P.H.	I.A.S. KNOTS	U.S. G.P.H.	IMP. G.P.H.	R.P.M.	M.P. IN.HG.	I.A.S. M.P.H.	I.A.S. KNOTS	U.S. G.P.H.	IMP. G.P.H.	R.P.M.	M.P. IN.HG.	I.A.S. M.P.H.	I.A.S. KNOTS	U.S. G.P.H.	IMP. G.P.H.		R.P.M.	M.P. IN.HG.	I.A.S. M.P.H.	I.A.S. KNOTS	U.S. G.P.H.	IMP. G.P.H.
						30000																			30000						
						25000																			25000						
						20000																			20000						
						15000																			15000						
						12000																			12000						
2400	F.T.	180	156	165	137	9000							2300	32.5	145	126	150	125							9000						
2400	38	195	169	185	154	6000							2250	34	155	134	155	129	2100	31.5	135	117	115	96	6000						
2400	38	190	165	175	146	3000							2250	34.5	165	143	155	129	2100	32.5	145	126	120	100	3000	2100	32	130	113	100	83
2400	38	185	160	170	142	S.L.							2250	36	170	147	155	129	2100	33.5	155	134	125	104	S.L.	2100	32	140	121	100	83

LEGEND

1 Range values at 6,000 ft. only

2 Allow __________ U. S. gals., __________ Imp. gals. for warm up, take-off to __________ feet altitude and climb. Use fuel from tanks in the following order __________. Return fuel flows to tank

REFER TO "SPECIFIC ENGINE FLIGHT CHART" FOR ADDITIONAL ENGINE OPERATION DATA.

BOLD NUMBERS: Use Auto-Rich
LIGHT ITALICS: Use Auto-Lean
WITH TWO SPEED BLOWER: Use high blower above heavy line only

I.A.S.: Indicated Air Speed
T.A.S.: True Air Speed
M.P.: Manifold Pressure (In. Hg.)
U.S.G.P.H.: U. S. Gallons Per Hour
IMP.G.P.H.: Imperial Gallons Per Hour
F.T.: Full Throttle

MODEL (S)
B-25 C
B-25 D

FLIGHT OPERATION INSTRUCTION CHART
(SINGLE ENGINE OPERATION)

GR. WT. ...... 26,000 ...... TO ...... 24,000 ...... POUNDS

SHEET ... 8 ... OF ... 8 ... SHEETS

EXTERNAL LOAD ITEMS
NONE

CONDITION	R.P.M.	M.P. (IN HG.)	BLOWER POSITION	MIXTURE POSITION	DURATION IN. MIN.
TAKE-OFF	2600	44.0	LOW	FULL RICH	5
MILITARY POWER	2600	42 / 41	LOW / HIGH	FULL RICH	5

ENGINE (S) R-2600-13 (1)

INSTRUCTIONS FOR USING CHART: Select figure in fuel column equal to or less than total amount of fuel in airplane. Move horizontally to the right or left and select a figure equal to or greater than the air miles to be flown. Vertically below and opposite desired cruising altitude read optimum cruising conditions. **NOTES:** (A) Avoid continuous cruising in Column I except in emergency. (B) Columns (II, III, IV & V) toward the right progressively give increase in range at sacrifice in speed. (C) Manifold Pressure (M.P.), Gallons Per Hour (G.P.H.), and True Airspeed are approximate maximum values for reference. (D) For quick reference, take-off and military power data are listed in the upper left corner of chart.

(NO WIND) — ALTERNATE CRUISING CONDITIONS — (NO RESERVE FUEL ALLOWANCE)

I (MAX. CONT. POWER) RANGE IN AIR MILES		FUEL U.S.GALS. 2	II RANGE IN AIR MILES		III RANGE IN AIR MILES		IV RANGE IN AIR MILES		FUEL IMP. GALS. 2	V (MAX. RANGE) RANGE IN AIR MILES	
STATUTE	NAUTICAL		STATUTE	NAUTICAL	STATUTE	NAUTICAL	STATUTE	NAUTICAL		STATUTE	NAUTICAL
540	470	500	620	540	700	610	790	690	417	880	770
430	370	400	490	430	560	490	630	550	333	710	620
320	280	300	370	320	420	360	470	410	250	530	460
210	180	200	250	210	280	240	310	270	167	350	300
100	100	100	120	110	140	120	160	130	83	170	150

OPERATING DATA — Columns I, II, III

DENSITY ALT. IN FEET	I R.P.M.	I M.P. IN.HG.	I T.A.S. M.P.H.	I T.A.S. KNOTS	I U.S. G.P.H.	I IMP. G.P.H.	II R.P.M.	II M.P. IN.HG.	II I.A.S. M.P.H.	II I.A.S. KNOTS	II U.S. G.P.H.	II IMP. G.P.H.	III R.P.M.	III M.P. IN.HG.	III I.A.S. M.P.H.	III I.A.S. KNOTS	III U.S. G.P.H.	III IMP. G.P.H.
30000																		
25000																		
20000																		
15000																		
12000																		
9000	2400	F.T.	205	178	165	137	2300	30.5	160	138	150	125	2100	32.5	145	126	120	100
6000	2400	38	200	173	185	154	2200	33.5	165	143	150	125	2100	31	150	130	120	100
3000	2400	38	195	169	175	146	2200	34.5	170	147	150	125	2100	32.5	160	139	120	100
S. L.	2400	38	185	160	170	142	2200	35	170	147	145	121	2100	33.5	160	139	120	100

OPERATING DATA — Columns IV, V

DENSITY ALT. IN FEET	IV R.P.M.	IV M.P. IN.HG.	IV I.A.S. M.P.H.	IV I.A.S. KNOTS	IV U.S. G.P.H.	IV IMP. G.P.H.	V R.P.M.	V M.P. IN.HG.	V I.A.S. M.P.H.	V I.A.S. KNOTS	V U.S. G.P.H.	V IMP. G.P.H.
30000												
25000												
20000												
15000												
12000												
9000												
6000	2050	30	135	117	95	79						
3000	2050	30	150	130	100	83	2000	29.5	135	117	80	67
S. L.	2050	32	155	134	100	83	2000	30.5	140	121	80	67

LEGEND

1 Range values at 6,000 ft. only

2 Allow _____ U. S. gals., _____ Imp. gals. for warm up, take-off to _____ feet altitude and climb. Use fuel from tanks in the following order _____ Return fuel flows to tank

BOLD NUMBERS: Use Auto-Rich
LIGHT ITALICS: Use Auto-Lean
WITH TWO SPEED BLOWER: Use high blower above heavy line only

I.A.S.: Indicated Air Speed
T.A.S.: True Air Speed
M.P.: Manifold Pressure (In. Hg.)
U.S.G.P.H.: U. S. Gallons Per Hour
IMP.G.P.H.: Imperial Gallons Per Hour
F.T.: Full Throttle

REFER TO "SPECIFIC ENGINE FLIGHT CHART" FOR ADDITIONAL ENGINE OPERATION DATA.

5-1-42

AIRPLANE MODELS
B-25 C & B-25 D
NO EXTERNAL LOAD

TAKE-OFF, CLIMB & LANDING CHART
(SHEET 1 OF 3 SHEETS)

ENGINE MODELS
R-2600-13

TAKE-OFF DISTANCE (IN FEET)

GROSS WEIGHT (IN LBS.)	HEAD WIND (MPH)	HARD SURFACE RUNWAY						SOD-TURF RUNWAY						SOFT SURFACE RUNWAY					
		AT SEA LEVEL		AT 3,000 FT.		AT 6,000 FT.		AT SEA LEVEL		AT 3,000 FT.		AT 6,000 FT.		AT SEA LEVEL		AT 3,000 FT.		AT 6,000 FT.	
		GROUND RUN	TO CLEAR 50' OBJ.	GROUND RUN	TO CLEAR 50' OBJ.	GROUND RUN	TO CLEAR 50' OBJ.	GROUND RUN	TO CLEAR 50' OBJ.	GROUND RUN	TO CLEAR 50' OBJ.	GROUND RUN	TO CLEAR 50' OBJ.	GROUND RUN	TO CLEAR 50' OBJ.	GROUND RUN	TO CLEAR 50' OBJ.	GROUND RUN	TO CLEAR 50' OBJ.
36,000	0	4600	5700	5200	6400	6700	8100	5300	6300	6000	7200	7600	9200	8000	9000	9300	10,500	11,500	13,000
36,000	20	3200	4000	3400	4300	4500	5600	3500	4300	4000	5000	5400	6600	5300	6100	6300	7,300	7,800	9,000
36,000	40	2000	2700	2100	2800	2800	3700	2100	2900	2600	3400	3500	4600	3400	4000	4200	5,000	5,100	6,100
32,000	0	3300	4200	3700	4700	4700	5900	3700	4600	4000	5000	5300	6600	5500	6300	5900	6,800	8,200	9,500
32,000	20	2200	2900	2500	3300	3200	4200	2500	3200	2800	3600	3400	4400	3700	4300	4000	4,800	5,800	6,800
32,000	40	1400	2000	1600	2300	2000	2800	1700	2200	1800	2400	2100	2900	2300	2800	2700	3,300	3,900	4,700
28,000	0	2300	3000	2500	3300	3000	4000	2500	3300	2700	3500	3200	4200	3500	4200	3700	4,500	4,800	5,800
28,000	20	1600	2200	1700	2400	2100	2900	1700	2400	1800	2500	2100	3000	2300	2900	2500	3,100	3,300	4,200
28,000	40	1000	1500	1100	1600	1300	2000	1100	1600	1200	1800	1400	2100	1500	2000	1600	2,100	2,000	2,600

NOTE: INCREASE DISTANCE 10% FOR EACH 10°C (50°F) ABOVE 0°C (32°F)   ENGINE LIMITS FOR TAKE-OFF   2600   RPM &   44.0   IN. HG

CLIMB DATA

COMBAT MISSIONS USE   2400   RPM &   38   IN. HG   FERRY MISSIONS USE   2100   RPM &   31.5   IN. HG

GROSS WEIGHT (IN LBS.)	TYPE OF CLIMB	S.L. TO 3000 FT. ALT.			6000 FT. ALT.				9000 FT. ALT.				12,000 FT. ALT.				15,000 FT. ALT.				BLOWER CHANGE
		BEST I.A.S.	FT./MIN.	TIME FROM S.L.	BEST I.A.S.	FT./MIN.	TIME FROM S.L.	FUEL FROM S.L.	BEST I.A.S.	FT./MIN.	TIME FROM S.L.	FUEL FROM S.L.	BEST I.A.S.	FT./MIN.	TIME FROM S.L.	FUEL FROM S.L.	BEST I.A.S.	FT./MIN.	TIME FROM S.L.	FUEL FROM S.L.	
36,000	COMBAT	145	1430	2:0	145	1380	4:5	35	145	1200	6:5	50	145	1040	9:5	65	140	880	12:5	80	11,000
36,000	FERRY	145	640	5:0	145	660	9:5	40	145	670	14:5	60	145	420	20:5	80	140	300	29:5	105	13,000
32,000	COMBAT	150	1560	2:0	150	1520	4:0	35	150	1340	6:0	45	150	1180	8:5	60	145	1630	11:0	75	11,000
32,000	FERRY	145	900	3:5	145	930	6:5	30	145	950	10:0	45	145	670	14:0	60	140	550	19:0	75	13,000
28,000	COMBAT	155	1750	2:0	155	1640	3:5	30	155	1480	5:5	40	155	1340	8:0	55	150	1190	10:0	70	11,000
28,000	FERRY	145	1180	3:0	145	1200	5:0	25	145	1230	7:0	35	145	950	11:0	45	140	830	13:0	55	13,000

NOTE: INCREASED ELAPSED CLIMBING TIME   % FOR EACH 10°C (50°F) ABOVE 0°C (32°F) FREE AIR TEMPERATURE.   FUEL INCLUDES WARM-UP AND TAKE-OFF ALLOWANCE

LANDING DISTANCE (IN FEET)

GROSS WEIGHT (IN LBS.)	BEST I.A.S. Approach	HARD DRY SURFACE						FIRM DRY SOD						WET OR SLIPPERY					
		AT SEA LEVEL		AT 3,000 FT.		AT 6,000 FT.		AT SEA LEVEL		AT 3,000 FT.		AT 6,000 FT.		AT SEA LEVEL		AT 3,000 FT.		AT 6,000 FT.	
		TO CLEAR 50' OBJ.	GROUND ROLL	TO CLEAR 50' OBJ.	GROUND ROLL	TO CLEAR 50' OBJ.	GROUND ROLL	TO CLEAR 50' OBJ.	GROUND ROLL	TO CLEAR 50' OBJ.	GROUND ROLL	TO CLEAR 50' OBJ.	GROUND ROLL	TO CLEAR 50' OBJ.	GROUND ROLL	TO CLEAR 50' OBJ.	GROUND ROLL	TO CLEAR 50' OBJ.	GROUND ROLL
34,000	120	3900	1900	4100	2100	4400	2300	4200	2200	4400	2400	4800	2600	9300	7400	10,200	8200	11,100	8900
31,000	120	3600	1700	3800	1900	4100	2100	3300	2000	4100	2200	4400	2400	8700	6800	9,400	7800	10,200	8200
25,000	110	3100	1400	3300	1500	3500	1700	3000	1600	3500	1800	3800	1900	7200	5500	7,800	6000	8,400	6600

NOTE: FOR GROUND TEMPERATURES ABOVE 35°C (95°F) INCREASE APPROACH I.A.S. 10% AND ALLOW 20% INCREASE IN GROUND ROLL.

REMARKS

LEGEND
I. A. S.: Indicated Air Speed

NOTE: All distances are average, and subject to considerable variations because of differences in pilot technique, load, C.G., etc.

RED FIGURES HAVE NOT BEEN FLIGHT CHECKED.

TAKE-OFF, CLIMB & LANDING CHART
(SHEET 2 OF 3 SHEETS)

AIRPLANE MODELS B-25 C & B-25 D WITH TORPEDO OR 8 WING BOMBS

ENGINE MODELS R-2600-13

5-1-42

TAKE-OFF DISTANCE (IN FEET)

GROSS WEIGHT (IN LBS.)	HEAD WIND (MPH)	HARD SURFACE RUNWAY						SOD-TURF RUNWAY						SOFT SURFACE RUNWAY					
		AT SEA LEVEL		AT 3,000 FT.		AT 6,000 FT.		AT SEA LEVEL		AT 3,000 FT.		AT 6,000 FT.		AT SEA LEVEL		AT 3,000 FT.		AT 6,000 FT.	
		GROUND RUN	TO CLEAR 50' OBJ.	GROUND RUN	TO CLEAR 50' OBJ.	GROUND RUN	TO CLEAR 50' OBJ.	GROUND RUN	TO CLEAR 50' OBJ.	GROUND RUN	TO CLEAR 50' OBJ.	GROUND RUN	TO CLEAR 50' OBJ.	GROUND RUN	TO CLEAR 50' OBJ.	GROUND RUN	TO CLEAR 50' OBJ.	GROUND RUN	TO CLEAR 50' OBJ.
36,000	0	4600	5700	5200	6400	6700	8100	5300	6300	6000	7200	7600	9200	8000	9000	9300	10,500	11,500	13,000
36,000	20	3200	4000	3400	4300	4500	5600	3500	4300	4000	5000	5400	6600	5300	6100	6300	7,300	7,800	9,000
36,000	40	2000	2700	2100	2800	2800	3700	2100	2900	2600	3400	3500	4600	3400	4000	4200	5,000	5,100	6,100
32,000	0	3300	4200	3700	4700	4700	5900	3700	4600	4000	5000	5300	6600	5500	6300	5900	6,800	8,200	9,500
32,000	20	2200	2900	2500	3300	3200	4200	2500	3200	2800	3600	3400	4400	3700	4300	4000	4,800	5,800	6,800
32,000	40	1400	2000	1600	2300	2000	2800	1700	2200	1800	2400	2100	2900	2300	2800	2700	3,300	3,900	4,700
28,000	0	2300	3000	2500	3300	3000	4000	2500	3300	2700	3500	3200	4200	3500	4200	3700	4,500	4,800	5,800
28,000	20	1600	2200	1700	2400	2100	2900	1700	2400	1800	2500	2100	3000	2300	2900	2500	3,100	3,300	4,200
28,000	40	1000	1500	1100	1600	1300	2000	1100	1600	1200	1800	1400	2100	1500	2000	1600	2,100	2,000	2,600

NOTE: INCREASE DISTANCE 10% FOR EACH 10°C (50°F) ABOVE 0°C (32°F)

ENGINE LIMITS FOR TAKE-OFF 2600 RPM & 44.0 IN. HG

CLIMB DATA

COMBAT MISSIONS USE 2400 RPM & 38.0 IN. HG

FERRY MISSIONS USE 2100 RPM & 31.5 IN. HG

GROSS WEIGHT (IN LBS.)	TYPE OF CLIMB	S.L. TO 3000 FT. ALT.			6000 FT. ALT.				9000 FT. ALT.				12,000 FT. ALT.				15,000 FT. ALT.				BLOWER CHANGE
		BEST I.A.S.	FT./MIN.	TIME FROM S.L.	BEST I.A.S.	FT./MIN.	TIME FROM S.L.	FUEL FROM S.L.	BEST I.A.S.	FT./MIN.	TIME FROM S.L.	FUEL FROM S.L.	BEST I.A.S.	FT./MIN.	TIME FROM S.L.	FUEL FROM S.L.	BEST I.A.S.	FT./MIN.	TIME FROM S.L.	FUEL FROM S.L.	
36,000	COMBAT	145	1050	3:5	145	1020	5:5	40	145	830	9:0	60	145	650	13:0	80	140	500	18:0	115	11,000
	FERRY	145	420	7:0	145	430	14:0	55	145	430	21:0	80	145	180	31:0	110	140	50	59:5	200	13,000
32,000	COMBAT	150	1180	3:0	150	1170	5:0	40	150	960	8:0	55	150	790	11:0	75	145	640	15:5	100	11,000
	FERRY	145	660	4:5	145	670	9:0	40	145	680	13:5	55	145	400	19:0	75	140	280	28:0	100	13,000
28,000	COMBAT	155	1310	2:5	155	1300	4:5	35	155	1100	7:5	50	155	930	10:5	65	150	790	14:0	85	11,000
	FERRY	145	940	3:5	145	960	6:5	30	145	980	9:5	40	145	680	13:5	55	140	560	18:0	70	13,000

NOTE: INCREASED ELAPSED CLIMBING TIME 10% FOR EACH 10°C (50°F) ABOVE 0°C (32°F) FREE AIR TEMPERATURE.

FUEL INCLUDES WARM-UP AND TAKE-OFF ALLOWANCE

LANDING DISTANCE (IN FEET)

GROSS WEIGHT (IN LBS.)	BEST I.A.S. Approach	HARD DRY SURFACE						FIRM DRY SOD						WET OR SLIPPERY					
		AT SEA LEVEL		AT 3,000 FT.		AT 6,000 FT.		AT SEA LEVEL		AT 3,000 FT.		AT 6,000 FT.		AT SEA LEVEL		AT 3,000 FT.		AT 6,000 FT.	
		TO CLEAR 50' OBJ.	GROUND ROLL	TO CLEAR 50' OBJ.	GROUND ROLL	TO CLEAR 50' OBJ.	GROUND ROLL	TO CLEAR 50' OBJ.	GROUND ROLL	TO CLEAR 50' OBJ.	GROUND ROLL	TO CLEAR 50' OBJ.	GROUND ROLL	TO CLEAR 50' OBJ.	GROUND ROLL	TO CLEAR 50' OBJ.	GROUND ROLL	TO CLEAR 50' OBJ.	GROUND ROLL
34,000	120	3900	1900	4100	2100	4400	2300	4200	2200	4400	2400	4800	2600	9300	7400	10,200	8200	11,100	8900
31,000	120	3600	1700	3800	1900	4100	2100	3900	2000	4100	2200	4400	2400	8700	6800	9,400	7500	10,200	8200
25,000	110	3100	1400	3300	1500	3500	1700	3000	1600	3500	1800	3800	1900	7200	5500	7,800	6000	8,400	6600

NOTE: FOR GROUND TEMPERATURES ABOVE 35°C (95°F) INCREASE APPROACH I.A.S. 10% AND ALLOW 20% INCREASE IN GROUND ROLL.

REMARKS

LEGEND

I. A. S.: Indicated Air Speed

NOTE: All distances are average, and subject to considerable variations because of differences in pilot technique, load, C.G., etc.

RED FIGURES HAVE NOT BEEN FLIGHT CHECKED.

AIRPLANE MODELS
B-25 C & B-25 D
WITH TORPEDO AND 8 WING BOMBS

TAKE-OFF, CLIMB & LANDING CHART
(SHEET 3 OF 3 SHEETS)

ENGINE MODELS
R-2600-13

5-1-42

TAKE-OFF DISTANCE (IN FEET)

GROSS WEIGHT (IN LBS.)	HEAD WIND (MPH)	HARD SURFACE RUNWAY						SOD-TURF RUNWAY						SOFT SURFACE RUNWAY					
		AT SEA LEVEL		AT 3,000 FT.		AT 6,000 FT.		AT SEA LEVEL		AT 3,000 FT.		AT 6,000 FT.		AT SEA LEVEL		AT 3,000 FT.		AT 6,000 FT.	
		GROUND RUN	TO CLEAR 50' OBJ.	GROUND RUN	TO CLEAR 50' OBJ.	GROUND RUN	TO CLEAR 50' OBJ.	GROUND RUN	TO CLEAR 50' OBJ.	GROUND RUN	TO CLEAR 50' OBJ.	GROUND RUN	TO CLEAR 50' OBJ.	GROUND RUN	TO CLEAR 50' OBJ.	GROUND RUN	TO CLEAR 50' OBJ.	GROUND RUN	TO CLEAR 50' OBJ.
36,000	0	4600	5700	5200	6400	6700	8100	5300	6300	6000	7200	7600	9200	8000	9000	9300	10,500	11,500	13,000
36,000	20	3200	4000	3400	4300	4500	5600	3500	4300	4000	5000	5400	6600	5300	6100	6300	7,300	7,800	9,000
36,000	40	2000	2700	2100	2800	2800	3700	2100	2900	2600	3400	3500	4600	3400	4000	4200	5,000	5,100	6,100
32,000	0	3300	4200	3700	4700	4700	5900	3700	4600	4000	5000	5300	6600	5500	6300	5900	6,800	8,200	9,500
32,000	20	2200	2900	2500	3300	3200	4200	2500	3200	2800	3600	3400	4400	3700	4300	4000	4,800	5,800	6,800
32,000	40	1400	2000	1600	2300	2000	2800	1700	2200	1800	2400	2100	2900	2300	2800	2700	3,300	3,900	4,700
28,000	0	2300	3000	2500	3300	3000	4000	2500	3300	2700	3500	3200	4200	3500	4200	3700	4,500	4,800	5,800
28,000	20	1600	2200	1700	2400	2100	2900	1700	2400	1800	2500	2100	3000	2300	2900	2500	3,100	3,300	4,200
28,000	40	1000	1500	1100	1600	1300	2000	1100	1600	1200	1800	1400	2100	1500	2000	1600	2,100	2,000	2,600

NOTE: INCREASE DISTANCE 10% FOR EACH 10°C (50°F) ABOVE 0°C (32°F)   ENGINE LIMITS FOR TAKE-OFF   2600   RPM &   44.0   IN. HG

CLIMB DATA

COMBAT MISSIONS USE   2400   RPM &   38   IN. HG   FERRY MISSIONS USE   2100   RPM &   31.5   IN. HG

GROSS WEIGHT (IN LBS.)	TYPE OF CLIMB	S.L. TO 3000 FT. ALT.			6000 FT. ALT.				9000 FT. ALT.				12,000 FT. ALT.				15,000 FT. ALT.				BLOWER CHANGE
		BEST I.A.S.	FT./MIN.	TIME FROM S.L.	BEST I.A.S.	FT./MIN.	TIME FROM S.L.	FUEL FROM S.L.	BEST I.A.S.	FT./MIN.	TIME FROM S.L.	FUEL FROM S.L.	BEST I.A.S.	FT./MIN.	TIME FROM S.L.	FUEL FROM S.L.	BEST I.A.S.	FT./MIN.	TIME FROM S.L.	FUEL FROM S.L.	
36,000	COMBAT	145	760	4:0	145	730	8:0	55	145	550	13:0	80	145	390	19:0	120	140	240	28:5	175	11,000
36,000	FERRY	140	320	9:5	140	320	19:0	70	140	320	28:5	105	140	60	47:5	160	-	-	-	-	13,000
32,000	COMBAT	150	890	3:5	150	870	6:5	50	150	680	10:5	70	150	530	15:5	100	145	390	22:0	140	11,000
32,000	FERRY	140	540	5:5	140	550	11:0	45	140	560	16:5	65	140	250	24:5	95	135	130	42:0	135	13,000
28,000	COMBAT	155	1020	3:0	155	1010	6:0	45	155	810	9:0	60	155	660	13:0	85	150	540	18:0	115	11,000
28,000	FERRY	140	830	4:0	140	850	7:5	35	140	860	11:0	45	140	550	15:5	60	135	410	21:0	80	13,000

NOTE: INCREASED ELAPSED CLIMBING TIME 10% FOR EACH 10°C (50°F) ABOVE 0°C (32°F) FREE AIR TEMPERATURE.   FUEL INCLUDES WARM-UP AND TAKE-OFF ALLOWANCE

LANDING DISTANCE (IN FEET)

GROSS WEIGHT (IN LBS.)	BEST I.A.S. Approach	HARD DRY SURFACE						FIRM DRY SOD						WET OR SLIPPERY					
		AT SEA LEVEL		AT 3,000 FT.		AT 6,000 FT.		AT SEA LEVEL		AT 3,000 FT.		AT 6,000 FT.		AT SEA LEVEL		AT 3,000 FT.		AT 6,000 FT.	
		TO CLEAR 50' OBJ.	GROUND ROLL	TO CLEAR 50' OBJ.	GROUND ROLL	TO CLEAR 50' OBJ.	GROUND ROLL	TO CLEAR 50' OBJ.	GROUND ROLL	TO CLEAR 50' OBJ.	GROUND ROLL	TO CLEAR 50' OBJ.	GROUND ROLL	TO CLEAR 50' OBJ.	GROUND ROLL	TO CLEAR 50' OBJ.	GROUND ROLL	TO CLEAR 50' OBJ.	GROUND ROLL
34,000	120	3900	1900	4100	2100	4400	2300	4200	2200	4400	2400	4800	2600	9300	7400	10,200	8200	11,100	8900
31,000	120	3600	1700	3800	1900	4100	2100	3900	2000	4100	2200	4400	2400	8700	6800	9,400	7500	10,200	8200
25,000	110	3100	1400	3300	1500	3500	1700	3000	1600	3500	1800	3800	1900	7200	5500	7,800	6000	8,400	6600

NOTE: FOR GROUND TEMPERATURES ABOVE 35°C (95°F) INCREASE APPROACH I.A.S. 10% AND ALLOW 20% INCREASE IN GROUND ROLL.

REMARKS

LEGEND

I. A. S.: Indicated Air Speed

NOTE: All distances are average, and subject to considerable variations because of differences in pilot technique, load, C.G., etc.

RED FIGURES HAVE NOT BEEN FLIGHT CHECKED.

SPECIFIC ENGINE FLIGHT CHART

AIRPLANE MODELS
B-25 C
B-25 D

ENGINE MODELS
R-2600-13
R-2600-13

CONDITION	FUEL PRESSURE LB./SQ. IN.	OIL PRESSURE LB./SQ. IN.	OIL TEMP. °C	COOLANT TEMP. °C		
DESIRED	6-7	80-85		50-70		
MAXIMUM	7	90		95 (Climb) / 85		
MINIMUM	6	75				
IDLING	6-7	25				

SUPERCHARGER TYPE: TWO SPEED

MAX. PERMISSIBLE DIVING R.P.M.

CONDITION	ALLOWABLE OIL CONSUMPTION	
"MAX CONTINUOUS"	46.6 IMP. PT./HR.	28 U.S. QT./HR.
"ECONOMICAL MAX."	25 IMP. PT./HR.	15 U.S. QT./HR.
"MIN. SPECIFIC"	IMP. PT./HR.	U.S. QT./HR.
OIL GRADE: (S)	(W)	

FUEL OCTANE 100

OPERATING CONDITION	R.P.M.	MANIF. PRESS. (BOOST.)	HORSE POWER	CRITICAL ALTITUDE (FEET)	BLOWER	USE LOW BLOWER BELOW	MIXTURE CONTROL POSITION	FUEL FLOW (GAL./HR./ENG.) U.S.	FUEL FLOW (GAL./HR./ENG.) IMP.	MAXIMUM CYL. TEMP. °C	MAXIMUM CYL. TEMP. °F	MAXIMUM DURATION (MINUTES)	REMARKS
TAKE-OFF	2600	44.0	1700	SEA LEVEL	L	ALWAYS FT. ALT.	FULL RICH	215	179	260	500	5	
EMERGENCY MAXIMUM	2600	42	1700	4,500	L	10,500	FULL	215	179	260	500	5	
		41	1700	12,000	H	FT. ALT.	RICH	196	163	260	500		
MAXIMUM CONTINUOUS	2400	38	1500	6,700	L	11,000	FULL	180	150	218	420		
		39	1350	15,000	H	FT. ALT.	RICH	205	171	205	400		
ECONOMICAL MAXIMUM	2100	31.5	1125	6,700	L	13,500	FULL	115	96	205	400		
		31.5	1012	15,000	H	FT. ALT.	RICH	115	96	205	400		
MINIMUM SPECIFIC CONSUMPTION	2100	29.5	1005	6,700	L	15,500	CRUISING	90	75	205	400		
		29.0	905	15,000	H	FT. ALT.	LEAN	100	83	205	400		
MINIMUM CRUISING	2000	27.0	900	6,700	L	16,500	CRUISING	75	63	205	400		
		27.0	810	15,000	H	FT. ALT.	LEAN	80	67	205	400		
CONDITIONS TO AVOID	1600	20.9	390	S.L.	L		CRUISING	31	26	205	400		Minimum Specific Fuel Flow
	1600	19.5	420	5,000	L			33	27.5	205	400		
	1600	18.5	450	10,000	L		LEAN	35	29	205	400		
	1600	18.0	490	15,000	L			37	31	205	400		

NOTE: CRITICAL ALTITUDE IS THAT AT WHICH MAXIMUM POWER IS OBTAINED WITH FULL THROTTLE UNDER CONDITIONS SHOWN.

5-1-42

SECTION IV

NAVIGATOR'S COMPARTMENT

1. GENERAL DESCRIPTION.

The navigator's compartment is located just aft of the pilot and copilot's seats. There is no bulkhead separating this compartment from the pilot's compartment. All the instruments and equipment necessary for the successful navigation of the airplane in flight are located in this compartment. In addition to the conventional equipment, the emergency nose wheel operating crank, pawl control and uplatch release are located at the forward right side of the compartment. The emergency bomb bay door control is also located in this compartment. The relief tube for the use of the bombardier, pilot and copilot is in this compartment. The front entrance hatch (figure 48) to be used by the bombardier, pilot, copilot and navigator is provided.

2. OPERATIONAL EQUIPMENT.

a. **The Emergency Nose Wheel Control Operation.**

(1) Ascertain from the pilot that the landing gear control handle (figure 21-192) is "DOWN."

(2) Pull nose gear emergency lock release (figure 25-238).

(3) Check with pilot or use drift meter to see that auxiliary gear is partially lowered before using lowering mechanism.

(4) Remove safety pin (figure 25-237) and turn pawl (figure 25-239) to "ON."

(5) Place crank (figure 25-241) on shaft and turn clockwise when facing until gear indicator registers (down and locked). If these instructions are violated the cables will break and the airplane will end in the repair shop.

CAUTION: Do not lower nose gear mechanically above an indicated air speed of 150 mph (130 knots). Do not return pawl to "OFF" until airplane is safely on the ground.

b. **Emergency Bomb Bay Door Control** (figure 22--204) Operation.

NOTE: Mechanical operation to be used only in the event of complete hydraulic pressure failure.

(1) Coordinate with the bombardier for correct position of his door operating and bomb release handle.

(2) Instruct bombardier to move bomb release handle (figure 35-314) to the doors "OPEN-RACK LOCK POSITION" or the "DOORS CLOSED" position as desired.

(3) Install bomb bay door operating crank on shaft in aft end of compartment. (Crank is stowed on lower right longeron opposite the crankshaft.) Install

the crank handle positioned downward. Turn crank clockwise to close doors and counterclockwise to open doors.

c. **Fuel Transfer System Controls.**

(1) To transfer fuel from fuel tanks on one side of airplane to forward wing fuel tanks on opposite side or to bomb bay droppable tank, or from the droppable tank to either of the forward wing fuel tanks, first connect the tanks desired by operating fuel transfer valves (figure 23-221). After setting the fuel transfer valves, move fuel flow direction switch (figure 23-229) in direction of desired flow and set transfer pump switch (figure 23-216) to "ON."

(2) The emergency fuel shut-off valves (figure 23-227) are located in the rear of the compartment.

d. **Heating and Ventilating Control.** - The air temperature control (figure 26-244) operates the heater and regulates its output.

e. **Electrical Controls.** - The generator main line switches (figure 23-223-226) and the generator voltage switch (figure 23-225) are to be turned to "OFF" in the event of a forced landing.

f. **Navigator Controls.**

(1) The radio compass control unit (figure 24-234) is located forward and below left window.

(2) The drift meter (figure 22-207) is located on right forward wall of compartment.

(3) **Drift Meter Controls.** - The electrically driven gyro must always be caged during take-off or landing, during turns when the incline is over 20 degrees from vertical when not in use, and before switching off the current. If the turn is started with the gyro uncaged, do not attempt to cage the gyro during the turn. Wait until the turn is completed and then cage the gyro. If the airplane inclines more than twenty degrees from the vertical in any direction during maneuvers, the gyro will strike its limit stops. This will cause the gyro to be moved from the vertical and will render all drift meter indications erroneous for several minutes after resuming level flight until the gyro has had time to erect itself. After an appreciable change of course, the gyro is apt to be slightly displaced from the vertical and a few minutes should be allowed for it to settle before taking observations.

(a) **Starting Operations.**

To start the electrically driven gyro of the drift meter turn on either of the battery disconnect switches and the gyro starting switch on the gyro housing.

(b) Allow the gyro to gain speed for at least 3 minutes.

(c) With the airplane in normal straight flight, uncage the gyro by pulling out the caging knob and moving it as far as it will go in the direction marked "UNCAGE." Make sure that caging knob snaps in place in the uncage position.

(d) Allow the gyro to run uncaged for at least ten minutes before attempting to take any readings.

(e) Turn the recticule light rheostat knob clockwise until the desired intensity of the grid lines is obtained.

g. <u>Automatic Flight Controls</u>. - A precessing switch is located below the right side window.

h. <u>Oxygen System</u>. - A type A-9A oxygen regulator (figure 22-205) is located on the right wall. To operate, attach hose to bayonet fitting and adjust regulator.

i. <u>Interphone System</u>.

(1) A jack box (figure 24-233) is located on the forward left wall. To operate, plug in ear phone cord and set switch to "CALL."

(2) An interphone amplifier is provided.

j. <u>Fluorescent Lighting System</u>.

(1) An inverter active and spare change over switch (figure 23-224) are located on the rear wall. If the active inverter fails, place the dual switches in the "SPARE" position.

<u>NOTE</u>: For normal operation, the switches will be kept in the "ACTIVE" position.

(2) A dome light switch (figure 27-247) and an extension light switch (figure 27-248) are provided and are operated conventionally.

k. <u>Fire Extinguisher</u>.

(1) One CO_2 fire extinguisher is located on the right side of the compartment.

(2) To operate, swing horn up, discharge close to base of fire and control with the trigger.

l. <u>Additional Equipment</u>.

(1) An anti-icer supply tank valve is located beneath tank in rear of the compartment.

(2) An oil strainer is located in lower right rear corner of the compartment.

(3) A drinking water container and cup holder is located at the left side of the compartment.

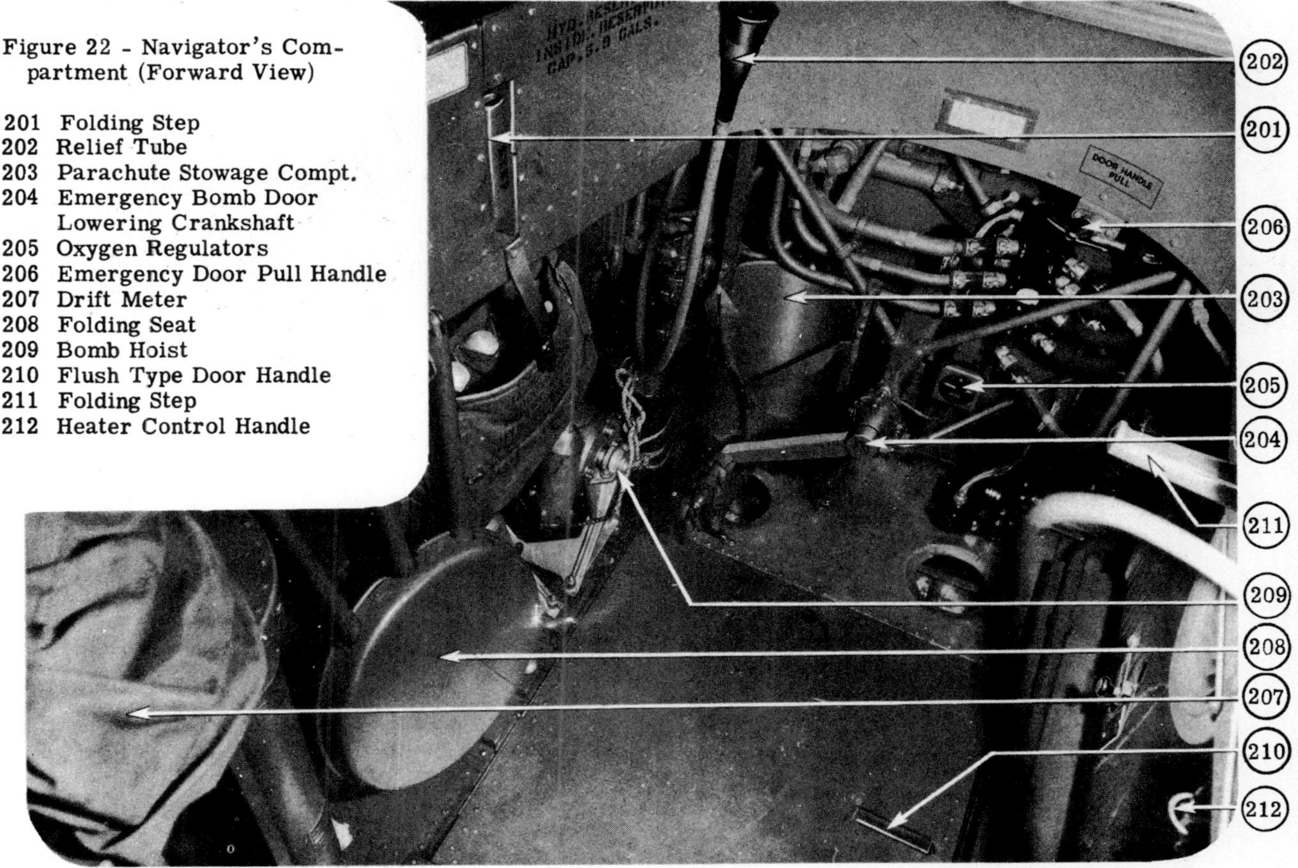

Figure 22 - Navigator's Compartment (Forward View)

201 Folding Step
202 Relief Tube
203 Parachute Stowage Compt.
204 Emergency Bomb Door Lowering Crankshaft
205 Oxygen Regulators
206 Emergency Door Pull Handle
207 Drift Meter
208 Folding Seat
209 Bomb Hoist
210 Flush Type Door Handle
211 Folding Step
212 Heater Control Handle

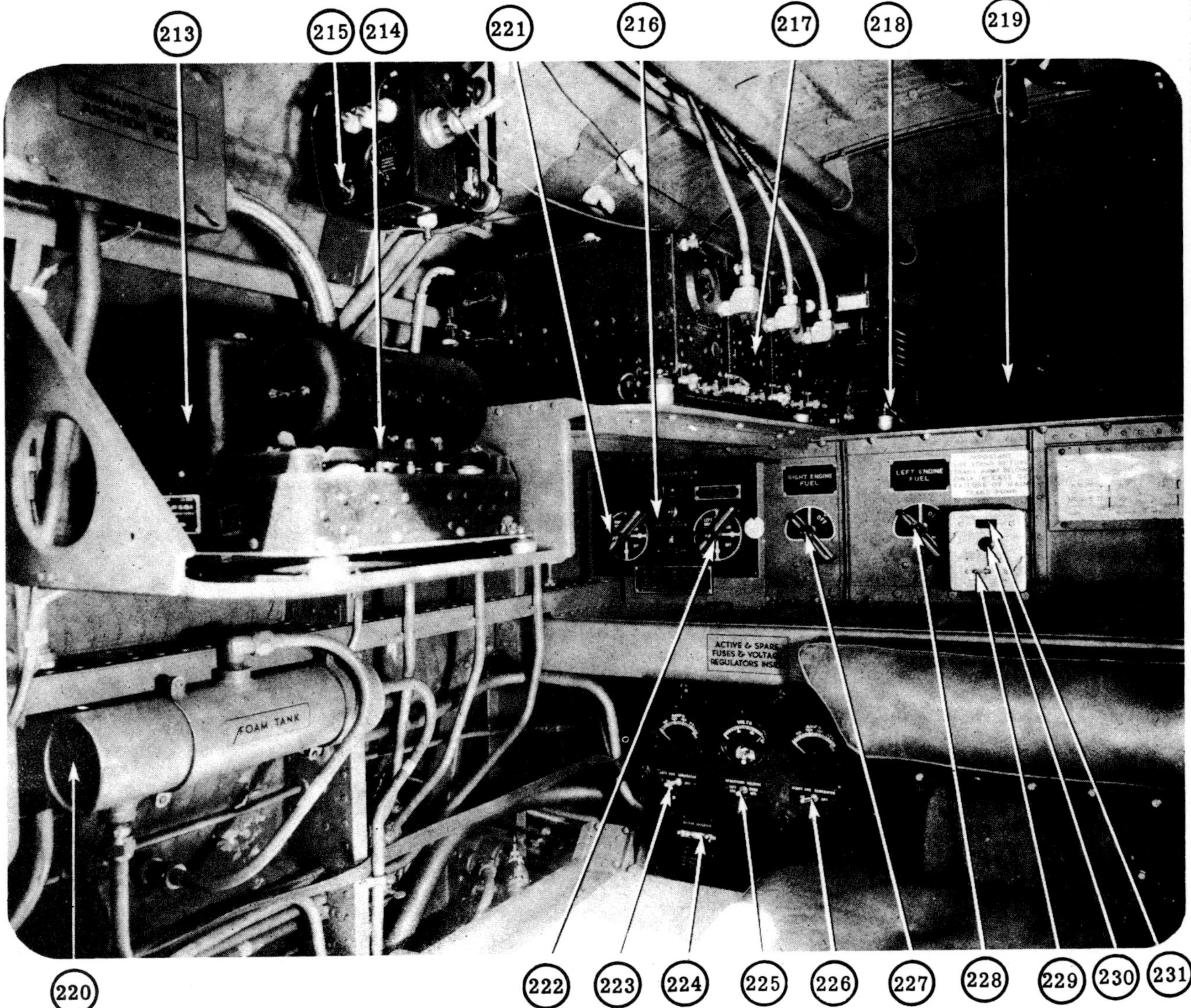

Figure 23 - Navigator's Compartment (Aft View)

213 Command Set Modulator
214 Dynamotor
215 Command Set Antenna Switch-
    ing Relay
216 Fuel Transfer Pump Switch
217 Command Set Receiver
218 Command Set Transmitter
219 Passageway Over Bomb Bay
220 Hydraulic System Foam Tank
221 Fuel Transfer Valve
222 Fuel Transfer Valve

223 Left Engine Generator Switch
224 Active Inverter Switch
225 Generator Voltage Switch
226 Right Engine Generator Switch
227 Emergency Fuel Shut-Off Valve
228 Emergency Spring Catch on
    Emergency Shut-Off Valve
229 Fuel Flow Direction Switch
230 Transfer Pump Indicator Light
231 Emergency Fuel Transfer
    Pump Switch

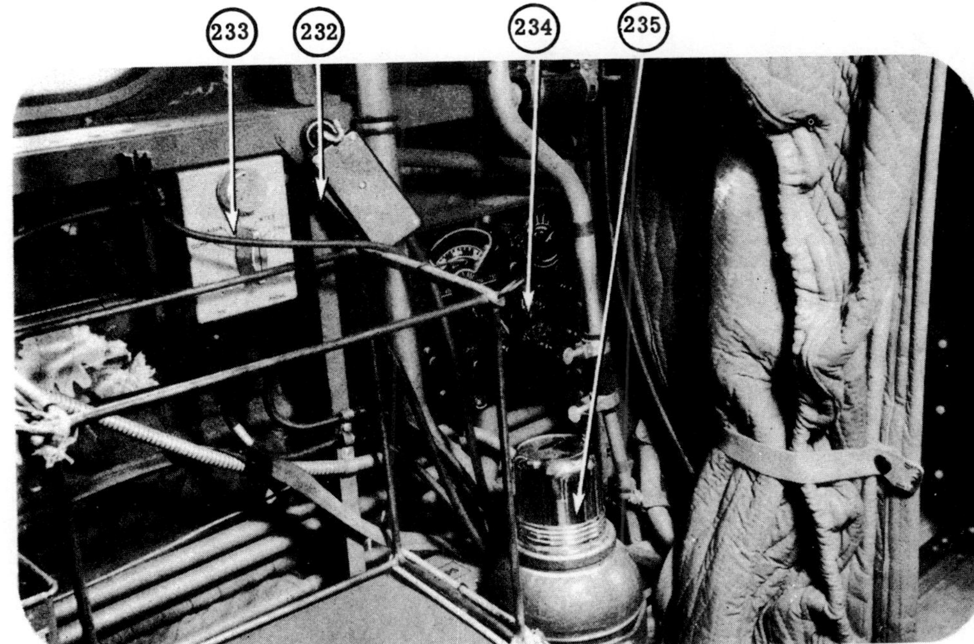

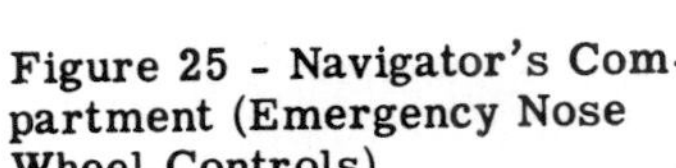

Figure 24 - Navigator's Compartment (Additional Equipment)

232 Push-To-Talk Switch
233 Jack Box
234 Radio Compass Control
235 Thermos Bottle

Figure 25 - Navigator's Compartment (Emergency Nose Wheel Controls)

236 Emergency Nose Gear Crankshaft
237 Safety Pin
238 Uplatch Release
239 Pawl Control
240 Emergency Air Brake Pressure Gage
241 Crank

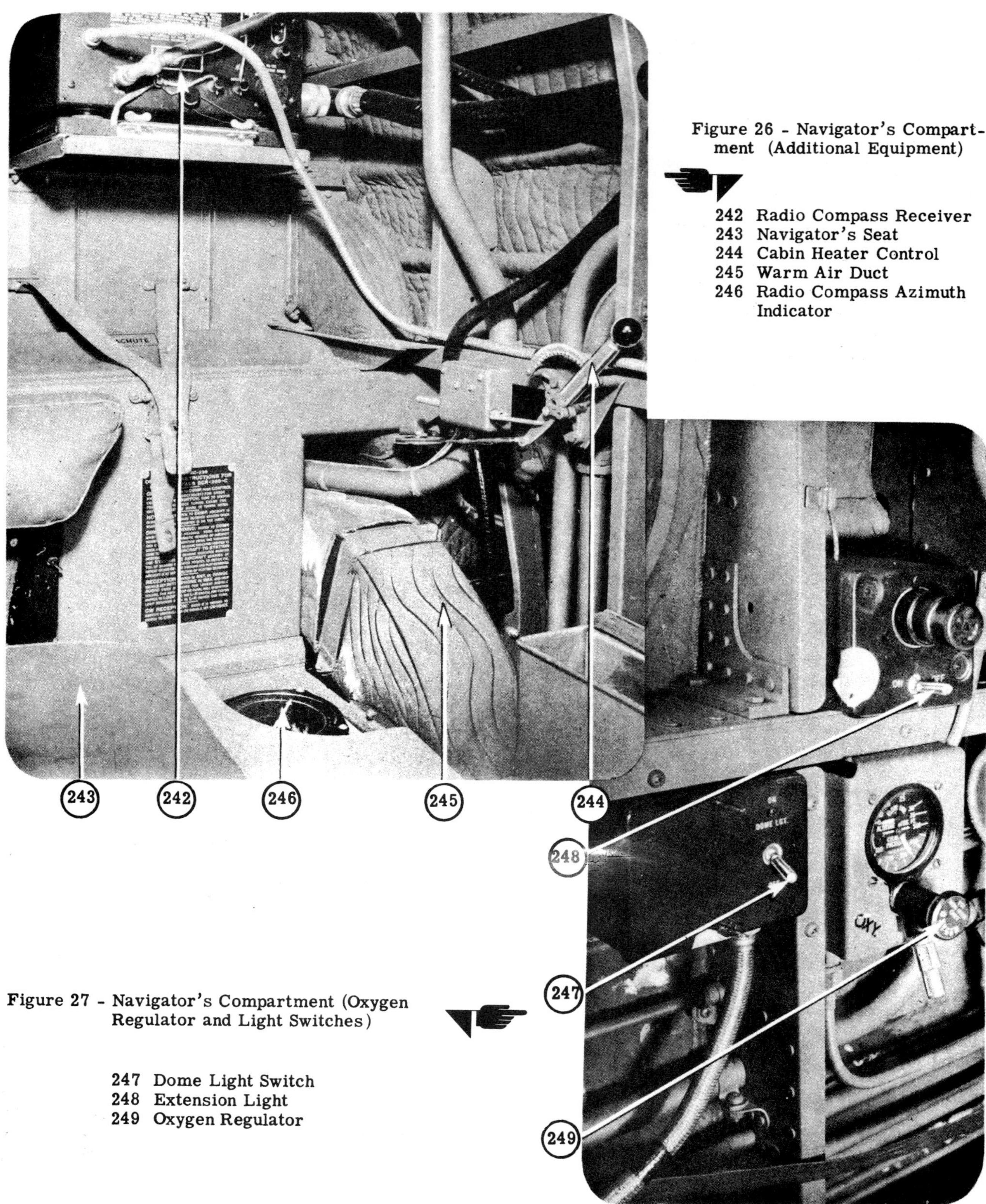

Figure 26 - Navigator's Compartment (Additional Equipment)

242 Radio Compass Receiver
243 Navigator's Seat
244 Cabin Heater Control
245 Warm Air Duct
246 Radio Compass Azimuth Indicator

Figure 27 - Navigator's Compartment (Oxygen Regulator and Light Switches)

247 Dome Light Switch
248 Extension Light
249 Oxygen Regulator

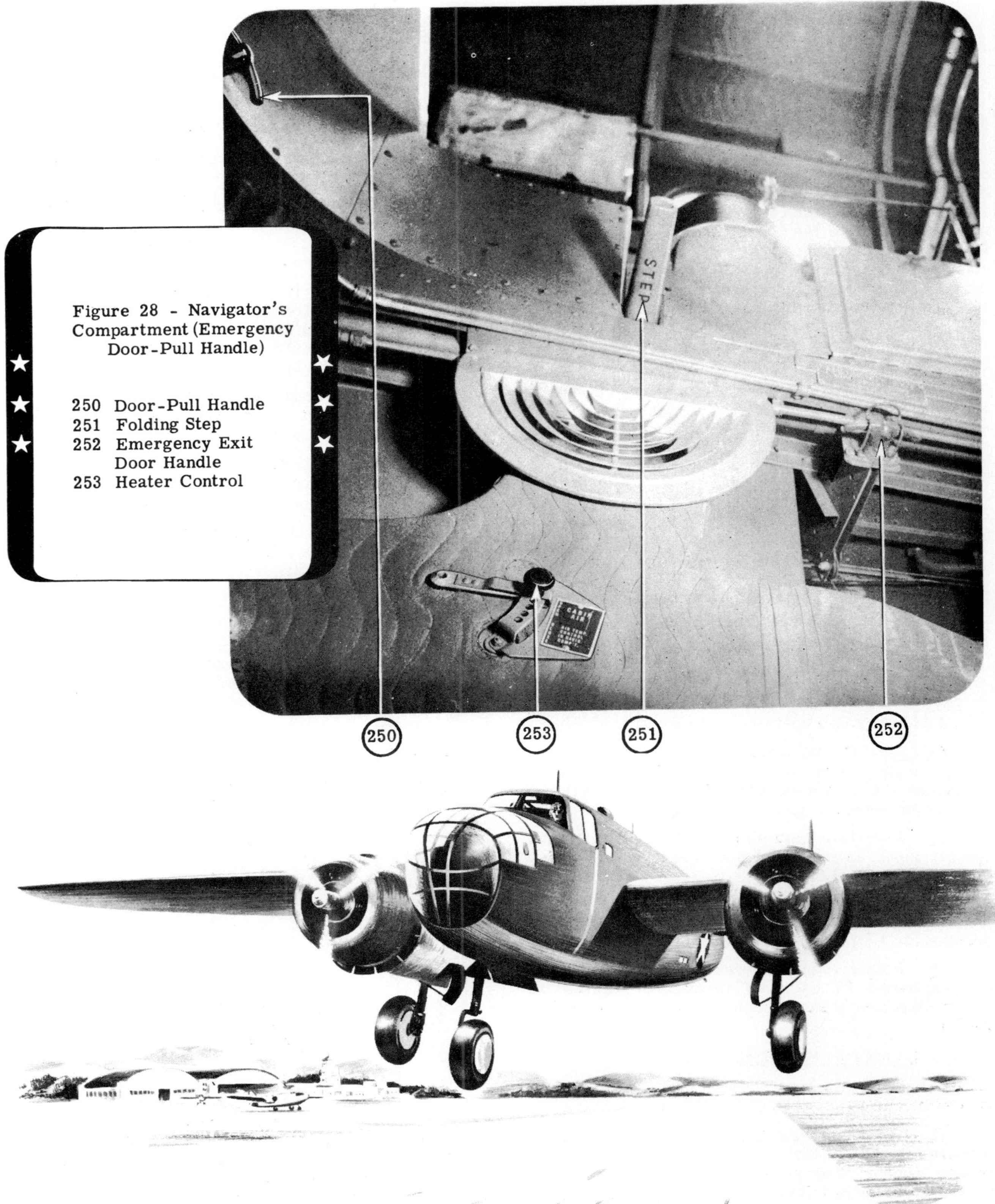

Figure 28 - Navigator's Compartment (Emergency Door-Pull Handle)

250 Door-Pull Handle
251 Folding Step
252 Emergency Exit Door Handle
253 Heater Control

SECTION V

RADIO OPERATOR AND LOWER GUN TURRET COMPARTMENT

1. GENERAL DESCRIPTION.

a. The compartment containing the radio equipment and the lower gun turret is located aft of the bomb bay and is just forward of the upper gun turret. It contains not only the radio transmitter and receiving sets, but also the lower model "J" gun turret. Provisions are made in this compartment for the emergency operation of the wing flaps and main landing gear. It is equipped with a data case, folding table, and a relief tube located below the operator's table. The relief tube is used by crew members stationed aft of the bomb bay. A side escape hatch (figure 14) is on the right side of the compartment.

b. The radio equipment and the lower gun turret are usually operated by the same man.

c. The communication equipment consists of the following units:

Command Set	SCR-274-N
Liaison Set, Medium Range	SCR-287-A
Radio Compass	SCR-269-A
Interphone Equipment	RC-36
Marker Beacon Receiving Equipment	RC-43

2. OPERATIONAL EQUIPMENT - RADIO OPERATOR'S COMPARTMENT.

a. Emergency Equipment.

(1) **Emergency Wing Flap Operation.**

NOTE: Interphone communication between pilot and the radio operator is necessary.

CAUTION: Do not operate the emergency flap operating system (do not even install crank) unless there is a complete failure of the hydraulic pressure system.

(a) Ascertain from pilot that his hydraulic flap control handle is at the extreme position for the desired movement of the flaps.

(b) Open hinged cover (figure 29-264) and install crank (figure 29-265) on shaft of flap operating mechanism.

(c) Rotate crank clockwise to raise the flaps (approximately 27 complete turns are required in either direction). Rotate counterclockwise to lower the flaps.

(2) **Emergency Operation of Main Landing Gear.**

(a) Ascertain that pilot's hydraulic landing gear control handle is set at the "DOWN" position.

(b) Disengage lower gun turret operator's chest support (figure 33-296) by pulling adjustment pin, raising support and allowing lower section of adjustment rod to fall forward. Hinge upper section of

adjustment rod and lower chest support.

(c) Lower radio operator's table (figure 29-262) to useful position.

(d) Release main landing gear operating screw assembly (held to forward wall of the compartment by finger-type fastener), rotate assembly away from wall until it locks in a fore and aft position.

(e) Pull main landing gear emergency up-position lock pin release (figure 29-267), located to the right of the lowering screw. This control releases the main gear up-position lock pins only and allows the main gear to partially lower, due to its own weight.

(f) Turn lowering screw handle clockwise until advised by pilot that his indicator registers that the main gear is down and locked.

(g) Ask pilot to retard throttle momentarily to ascertain that the landing gear is locked down as evidenced by the failure of the warning horn to sound.

WARNING: Improper operation of main landing gear emergency lowering screw may break cables and prove dangerous and cause expensive repair.

CAUTION: After lowering gear in the above manner, do not turn lowering screw handle counterclockwise to return handle to its original position on the wall of the compartment until the airplane is safely landed.

(<u>h</u>) After airplane is landed, turn main landing gear emergency lowering screw handle <u>counterclockwise</u>, returning handle to its original position and stow assembly.

(3) <u>Miscellaneous Emergency Equipment.</u>

(<u>a</u>) The side escape hatch is located on the right wall of the compartment.

(<u>b</u>) A CO_2 fire extinguisher is clipped to the right side wall of the compartment just aft of the escape hatch.

<u>b</u>. Communications Equipment.

(1) <u>Interphone Controls.</u> - An interphone jackbox is located at the aft end of the radio operator's table. The selector control has five selective positions as follows:

(<u>a</u>) "COMP". - The audio output of the compass receiver only is heard.

(<u>b</u>) "LIAISON". - The output of the liaison receiver and the sidetone of the liaison transmitter are heard. The microphone push-to-talk switch operates the transmit receive relay located within the liaison transmitter. The microphone will modulate the liaison transmitter when the switch is closed and the transmitter is in "VOICE" position.

NOTE: From the "LIAISON" position, voice transmission is effective only at the pilot's, copilot's and radio operator's station.

(<u>c</u>) "INTER". - At this position, all jack boxes provide intercommunication between crew members. The microphone connects to the input of the interphone amplifier, and the headphones to the output of the amplifier. In this position the volume control is not effective.

(<u>d</u>) "CALL". - This is an emergency call position in which all of the positions in all boxes are placed in parallel across the output of the amplifier. In an emergency any crew member desiring to call a station which is in use, may do so by switching to the "CALL" position. This position is effective on all stations.

(2) <u>Command Set.</u>

(<u>a</u>) The command receivers and transmitters of the SCR-274-N set are independently controlled. There are three markings on the transmitter switch: "VOICE", "CW", and "TONE." With the switch at the "VOICE" position, the microphone from any jack box switched to "COMMAND" is operative and the voice is transmitted when the push-to-talk button at that station is pressed. With the switch at the "CW" position, a continuous wave or an unmodulated signal is transmitted. With the switch at the "TONE" position, a signal is transmitted which is practically 100 percent modulated at 1000 cycles. For long range communication through interference "CW" is most effective, "TONE" next and "VOICE" least effective.

(<u>b</u>) At both the "CW" and "TONE" positions, the microphone is inoperative by voice, and signalling by code is accomplished by a key located on the top of the transmitter control box. An external or separate key may be used by plugging it into the jack marked "KEY." The push-to-talk button on the microphone may be used as a key for transmitting code when the control box switch is turned to the "CW" or "TONE" position. A microphone plugged into the jack marked "MIC" permits the operator to transmit voice over the command set only, instead of through the interphone system.

(3) <u>Receiver.</u> - The receiver control box is divided into three identical sections, each controlling the particular receiver to which it is electrically and mechanically connected. To receive a signal of specific frequency, use the section of the receiver control box which controls the particular receiver involved. The desired receiver is turned "ON" and "OFF" by a switch located in the upper right-hand corner of the control box section used. In addition to an "OFF" position, the switch has two selective positions marked "CW" and "MCW", each of which is an "ON" position and indicates the type of signal to be received.

(4) <u>Liaison Set Transmitter.</u>

(<u>a</u>) The "OFF"-"ON" switch of the liaison transmitter is located on the face of the transmitter case.

(<u>b</u>) With the transmitter turned "ON", the filament voltage, as indicated on the meter marked "FIL. VOLTAGE" must be within close limits of the line on the face of the meter at 10 volts. The "CW" and modulator filament voltages are checked individually by a switch adjacent to the "OFF"-"ON" switch. Each tuning unit contains tuned circuits to allow a transmitter output variable within the frequency limits specified on the tuning unit in use. A chart located on the face of each tuning unit indicates the dial settings for desired frequencies. It is possible with the monitor switch to tune the transmitter output frequency to match the frequency of any station coming in over the receiver. With the monitor switch at "MONITOR", the sidetone of the transmitter is automatically cut off and the receiver may be turned on with the "CW" switch and adjusted to a desired frequency as indicated on the dial, or to an incoming signal of a station which it is desired to contact. Then press the transmitter key and adjust the transmitter oscillator frequency dial until the transmitter frequency is heard in the receiver. This adjusts the transmitter and it will stand by ready for the break-in on the station to which it has been adjusted. With the monitor switch at "NORMAL" the receiver will be inoperative while the transmitter is operating.

(5) <u>Liaison Receiver.</u> - The liaison receiver is operated by a switch located on the front of the case and has three positions: "OFF", "MVC" and "AVC", the latter two designating manual volume control and automatic volume control, respectively. To tune in or search, turn the switch to "MVC" and upon obtaining the desired signal, change to "AVC." Operation

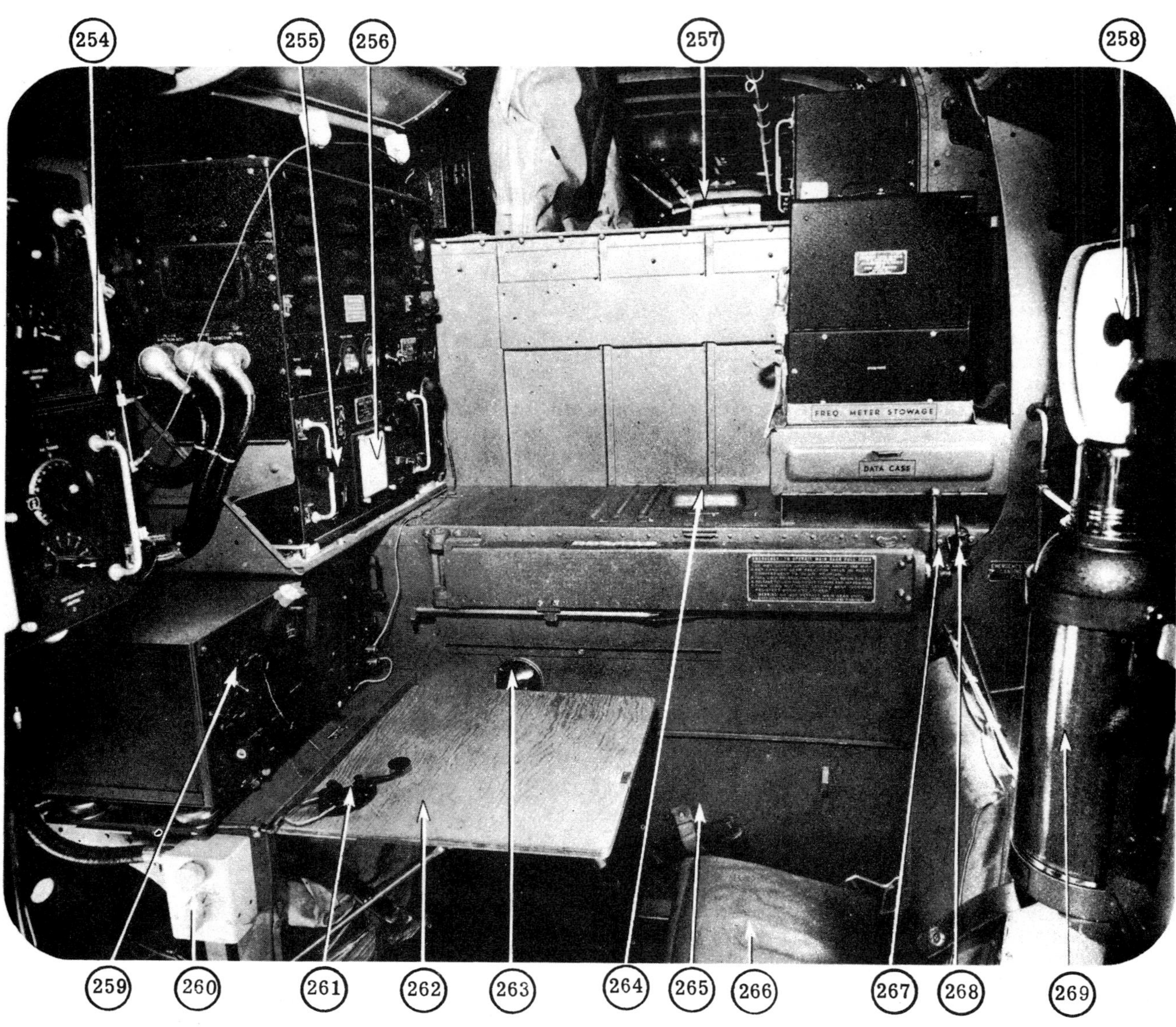

Figure 29 - Radio Operator's Compartment (Forward View)

254 Replaceable Transmitter Tuning Units
255 Hook for Holding Radio Operator's Table in Stowed Position
256 Transmitter
257 Passageway Over Bomb Bay
258 Oxygen Regulator
259 Receiver
260 Jack Box
261 Transmitting Key
262 Radio Operator's Table
263 Ash Receiver
264 Hinged Cover for Crankshaft for Emergency Flap Operation
265 Crank for Emergency Flap Operation
266 Radio Operator's Seat
267 Emergency Main Landing Gear Operating Screw
268 Up Position Lock Release for Main Landing Gear
269 Thermos Bottle

of the "BAND SWITCH" knob located on the face of the receiver case controls the selection of the frequency band. Dial calibrations, corresponding to the band selected, are revealed by the dial mask.

CAUTION: Do not make any adjustments within the command or liaison transmitter while high voltage supply is on. Do not remove or replace any tubes within any of the equipment while the equipment is turned on. Do not remove covers from any of the dynamotors or replace any fuses while dynamotors are operating. ■ ■ ■ ■ ■ ■ ►

(6) Trailing Antenna Control. - A reel control box(figure 41-352) is located on a bracket on the left-hand side of the radio operator's compartment immediately aft of the liaison receiver. The control knob may be turned from its central "OFF" position, right to "OUT", or left to "IN", in order to extend or retract the weighted trailing antenna. A three-digit visible counter indicates the number of turns made by the antenna in extending and reversing during retraction.

NOTE: An amber signal light above the counter will warn of an unsatisfactory landing con-

Figure 30 - Radio Operator's Compartment (Left Side)

270 Heater Control
271 Replaceable Transmitter Tuning Units
272 Replaceable Transmitter Tuning Units
273 Life Raft Stowage Compartment
274 Gun Sight Eye Cushion
275 Transmitter

dition if the trailing antenna is left in an extended position as the main landing gear starts to descend. <u>Do not</u> use the trailing antenna while on the ground.

(7) <u>Radio Compass</u>.

(<u>a</u>) The radio compass can be operated from either of the two control boxes, but not from both at the same time. The equipment is manually tuned from either remote control box, and electrical control is established at the desired control box by depressing the button marked "CONTROL" in the lower right-hand corner of the control box. When control is established at the desired remote control unit, a green indicating light will appear on the face of the unit.

(<u>b</u>) The radio compass performs the following functions:

<u>1</u>. Aural reception from the fixed antenna or from the rotatable loop. For signal reception during interference caused by precipitation, static or proximity of signals, the loop will prove superior.

<u>2</u>. Aural-null directional indication of an incoming signal with the loop only in use.

<u>3</u>. Visual uni-directional left-right indication of an incoming signal. The receiving unit is turned on or off by a switch on the face of the remote control box which in addition to having an "OFF" position marked thereon, has three other positions marked "COMP", "ANT", and "LOOP." With the switch turned to the position marked "COMP", both the rotatable loop and the fixed antenna are in use. In the "ANT" position, only the fixed antenna is in use and with the switch turned to "LOOP", only the rotatable loop is in use. Frequency band selection is accomplished by rotating the band switch handle to one of its three marked positions.

(8) <u>Marker Beacon</u>. - The operation of the marker beacon equipment is fully automatic. As the airplane passes over a fixed point from which a marker signal is being transmitted, the signal is picked up by the receiver, the output of which actuates a built-in relay. This relay in turn causes the indicator to flash on, thus indicating to the pilot that he has passed over a marker beacon.

Figure 31 - Radio Operator's Compartment (Oxygen Regulator and Fire Extinguisher)

276 Oxygen Regulator
277 Thermos Bottle
278 Horn of Fire Extinguisher
279 Fire Extinguisher (CO_2)

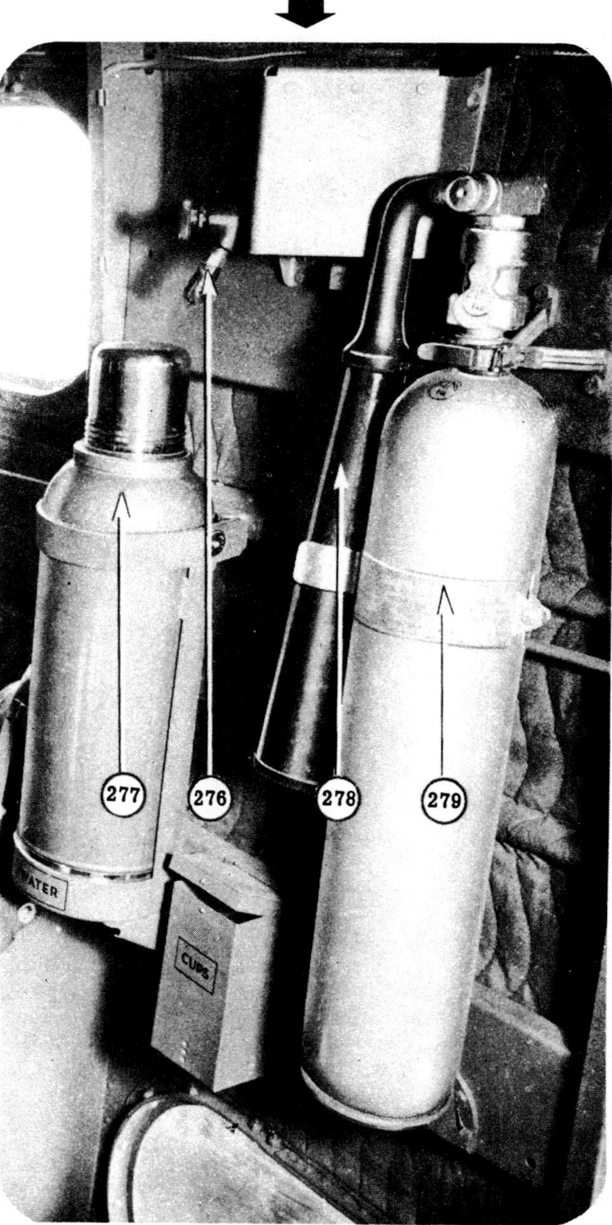

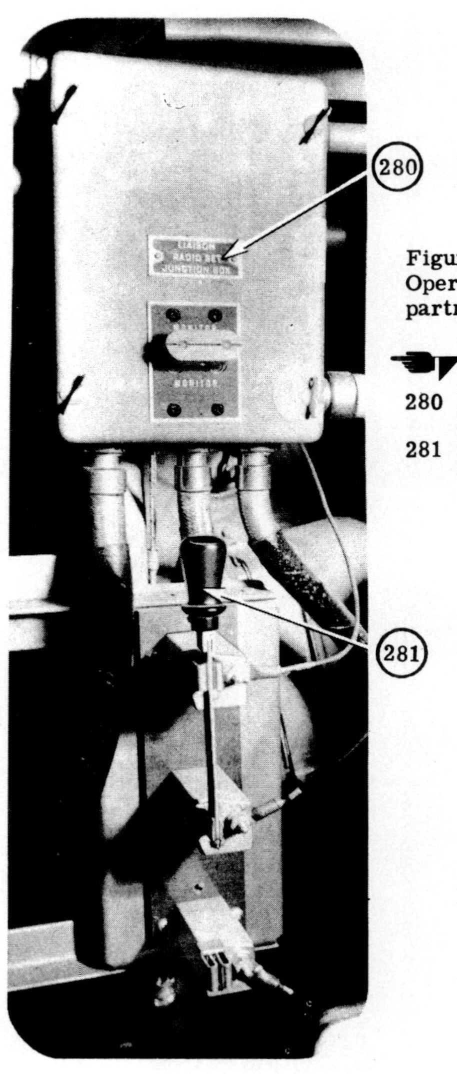

Figure 32 - Radio Operator's Compartment (Monitor Switch)

280 Liaison Set Junction Box
281 Monitor Switch

3. OPERATIONAL EQUIPMENT - LOWER GUN TURRET.

a. Gun Turret Operation.

CAUTION: Do not operate turret without adequate power supply. Damage to the turret or airplane may result if supply voltage at the turret is less than 20 volts.

(1) Extension to Combat Position.

(a) Firmly grasp control handle (figure 33-288) with right hand to depress master switch, but do not depress trigger switch.

(b) Turn control handle counterclockwise about vertical axis a few degrees for slow speed and maximum rotation for full speed.

CAUTION: As turret approaches extended position, use slow speed until indexing key auto-

matically engages and turret starts to rotate counterclockwise. Guns can be fired as soon as extended position is reached.

(2) Charging the Guns.

(a) To charge guns for combat or to remove faulty cartridge, depress control valve with knob rotated clockwise against stop (after depression, knob is automatically released at end of charging stroke, thereby positioning valve for next charging stroke).

(b) To charge guns and hold the bolt back in safety, depress control valve with knob rotated counterclockwise against stop. When ready to release for combat, rotate knob clockwise against the stop.

(3) Combat Operation.

(a) Lower the padded knee support well forward of turret.

(b) Adjust chest support as required by means of spring-loaded pin in chest support rod.

(c) To adjust focus of sight, pull up on knurled collar below sight cushion and rotate until desired focus is obtained. Then push collar down, engaging it in closest locking notch.

(d) Take kneeling position with right hand on control handle, left hand on steady grip and eye on sight eye cushion. (See figure 33-284.)

(e) Neutral position of control handle is halfway between vertical and horizontal rotation stops.

(f) To elevate the guns, rotate control handle vertically about its axis. The guns will move in same relative direction as handle, at a speed proportional to the degree of control handle rotation.

(g) To rotate turret in azimuth, rotate control handle horizontally about its vertical axis. Turret will move in same relative direction as handle, at a speed proportional to the degree of control handle rotation.

(h) Turn on windage indicator switch and rotate rheostat control to indicated air speed of aircraft determined from pilot. Microphone switch button is in the end of the steady grip.

(i) To fire guns, depress trigger firing switch in the control handle.

(j) Power is cut off by releasing grip on the control handle.

(4) Retraction to Stowed Position.

(a) Rotate turret slowly in clockwise direction with guns positioned approximately 12 degrees below horizontal. As guns approach aft position, firmly depress retraction lever, which will cause the turret to stop.

(b) Slowly raise guns in elevation, keeping control handle set for slow azimuth speed in clockwise direction, and retraction lever depressed. When guns

Figure 33 - Lower Gun Turret (Interior View)

282 Foot Support Adjustment	288 Steady Grip	ret Control
283 Upper Turret Shaft	289 Manual Turret Control Shifter	294 Turret Disconnect Switch
284 Gun Sight Eye Cushion	Shaft	295 Controller Box
285 Retract Lever	290 Ammunition Box	296 Turret Operator's Chest Support
286 Turret Rotation and Retraction	291 Spare Insert Tuning Units for	
Motor	Liaison Transmitter	297 Radio Operator's Table
287 Interphone Button - Inboard	292 Gun Sight Adjustment Collar	298 Turret Operator's Kneeling
End of Steady Grip	293 Crank Handle for Manual Tur-	Pad

enter index zone (2 to 6 degrees below horizontal), turret will again turn clockwise, index and start retracting.

CAUTION: Control handle must be kept at slow speed until turret begins to retract.

(c) After turret has retracted approximately one inch, release retracting lever (figure 33-285) and continue retraction at any desired speed to stowed position.

b. Emergency Turret Operation.

WARNING: Firing restrictor cams, turret control cams and gun control cams which are used to electrically stop the gunfire, turret rotation, and the gun elevation at predetermined positions, are also disengaged when operating turret mechanically. Therefore, prior to attempting to fire guns while operating turret mechanically, mechanically rotating turret or moving guns in elevation, make sure guns will not fire on airplane or strike fuselage.

(1) Remove required transparent sections on turret deck.

(2) Engage handcrank (figure 33-293) in manual gun control shaft and rotate crank as required to position the guns approximately 2 to 6 degrees below horizontal.

(3) Engage handcrank in manual turret control shaft (figure 33-289) pull shaft forward to disengage clutch from reduction gear mechanism.

(4) Rotate turret until guns are slightly to left of aft position. Firmly depress retraction lever and rotate turret clockwise.

(5) When aft position is reached, turret will begin to retract. Hold retraction lever depressed until turret has retracted approximately one inch, and then release lever. Retract turret to stowed position.

CAUTION: Do not raise turret beyond the point where the aft turret support brace contacts the retract-position limit switch located below the aft leg of the spider, as damage to the turret will result.

(6) Replace sections of transparent fairing on turret deck.

Figure 34 - Lower Gun Turret (Exterior View)

SECTION VI

BOMBARDIER'S COMPARTMENT

1. GENERAL DESCRIPTION.

The bombardier's compartment is located forward of and lower than the pilot's compartment. In the compartment, provision is made for the installation of the Norden M series, Estoppey D-8 or the British Mark IX bomb sights. The bomb controls provide for the "ARMED", "SAFE", "SELECTIVE", "TRAIN", "SALVO", and "EMERGENCY" releases. Ball and socket mounts (figure 36-322) for a .30 caliber machine gun are provided in the panels of the nose of this compartment. Stowage for the machine gun is by means of a socket and bracket at the right side of the compartment. Provision is also made for the installation of a type G-4 camera gun. In addition to the conventional equipment such as the interphone, ventilating and lighting systems this compartment has a side escape hatch (figure 14) to be used when the airplane is on the ground only. Armor plate on the seat and back of the bombardier's riding seat is the protection afforded in this compartment.

2. OPERATIONAL EQUIPMENT.

a. Bomb-Controls Operation.

(1) The bomb release handle cannot be removed from one setting to another until the plunger knob at the top of handle is pressed down. In order to move the bomb release handle from "SELECTIVE" to the "SALVO" position a mechanical safety guard must be hinged upward.

(2) When not in use, bomb controls shall be positioned as follows: bomb bay door control (figure 35-314) at the "DOORS CLOSED" position with anti-salvo guard in place; selective-train switches on "SELECTIVE" and the momentary contact bomb release switch with mechanical safety guard in place. After releasing bombs the controls shall be returned to the above positions to prevent accidental release of the remaining bombs.

(3) Setting controls prior to releasing bombs.

(a) In electrical selective or train release, the bombs are automatically electrically armed. In emergency salvo release, the bombs are always released automatically in the safe condition.

(b) Set the train selective switches to the desired position.

(c) If electrical train release of the bombs is desired, set the desired spacing on bomb release interval control. Do not move control while motor is running.

(d) To open bomb bay doors, move bomb release handle to "DOORS OPEN-RACK LOCK" position.

CAUTION: Do not open or close bomb bay doors on the ground without first verifying that all personnel are clear.

Be sure area under plane is clear before you open or close bomb doors!

(e) When door position indicator light illuminates, move bomb release handle to "SELECTIVE" position, if electrical train or selective release is desired.

(f) To close bomb bay doors return bomb release handle to "DOORS CLOSED" position.

(4) Selective Release. - Hinge bomb release switch safety guard (figure 35-317) upward and press release switch once for each bomb to be released.

(5) Train Release. - Hinge bomb release switch safety guard upward, and press and hold down release switch until the desired number of bombs has been released.

(6) Salvo Release. - Hinge anti-salvo guard upward and move bomb release handle (figure 35-314) to the extreme forward, "SALVO" position.

(7) Emergency Release by Pilot. - In the event the pilot has released the bombs or the droppable tank in an emergency, the bombardier performs the following operation.

(a) Move bomb release handle to the extreme forward end of the quadrant, past the "SALVO" po-

Figure 35 - Bombardier's Compartment (Left Side)

299 Free Air Temperature Gage
300 Bomb Sight Defroster Tube
  Stowed
301 Air-Speed Gage
302 Interval Control
303 Bomb Sight Defroster Tee
304 Interval Control
305 Selector Train Switch
306 Heater Control
307 Warning Kell
308 Warning Light
309 Pilot's Call Switch
310 Dome Light Switch
311 Cockpit Light Switch
312 Camera Power Switch
313 Identification Lights Switch
314 Bomb Release Handle
315 Salvo Switch
316 Bomb Rack Selector Switch
317 Bomb Release Switch

sition and then return handle to "DOORS CLOSED"
position. This will cock the pilot's emergency release unit and close doors.

 (b) The bombardier can also release the bombs
or the bomb bay droppable tank in an emergency by
using operation for "SALVO" release.

 CAUTION: If the hydraulic system has failed,
to release bombs from either the bombardier's
or pilot's compartment, the bomb bay doors
are to be cranked open from the navigator's
compartment. However, the bombardier's door
operating and bomb release handle must be
positioned.

 (8) Bombardier's Emergency Release. - Move
bomb release handle (figure 35-314) to the extreme
forward end of the quadrant, past the "SALVO" position and then return handle to the "SALVO" position.
If handle was in "SALVO" return handle to the "SE-

LECTIVE" position and then forward past "SALVO"
position in order to release bombs.

 b. Oxygen System.

 (1) Two type A-9A oxygen regulators are provided; one at bomb sight and one at riding seat.

 (2) To operate, attach hose to bayonet fitting and
adjust regulator.

 c. Interphone System.

 (1) A jack box (figure 37-336) is located at the
right side of the compartment.

 (2) To operate, plug in earphone cord and set
switch to desired point. A throat microphone switch
cord is also provided.

 d. Heating and Ventilating Equipment.

 (1) A controllable cold air scoop is provided on

the right side of the compartment.

(2) A valve equipped heat outlet (figure 36-324) is located on the left side of the compartment.

CAUTION: Do not open escape hatch during flight for ventilation.

e. Defrosting System.

(1) The bomb sight window receives heat whenever heater is on.

(2) A defroster tube (figure 35-303) for use on the bomb sight is provided. When not in use the tube may be stowed.

f. Automatic flight control equipment and pilot's call switch (figure 35-309) is located on bomb control panel.

g. Lighting System.

(1) A dome light switch (figure 35-310) is provided.

(2) An extension light switch (figure 36-319) is provided.

(3) Both of the above are operated conventionally.

h. Escape Hatch.

(1) A side escape hatch is located on the left side of the compartment.

(2) This hatch releases inboard.

CAUTION: This hatch is to be used when airplane is on the ground only and when motors are not running.

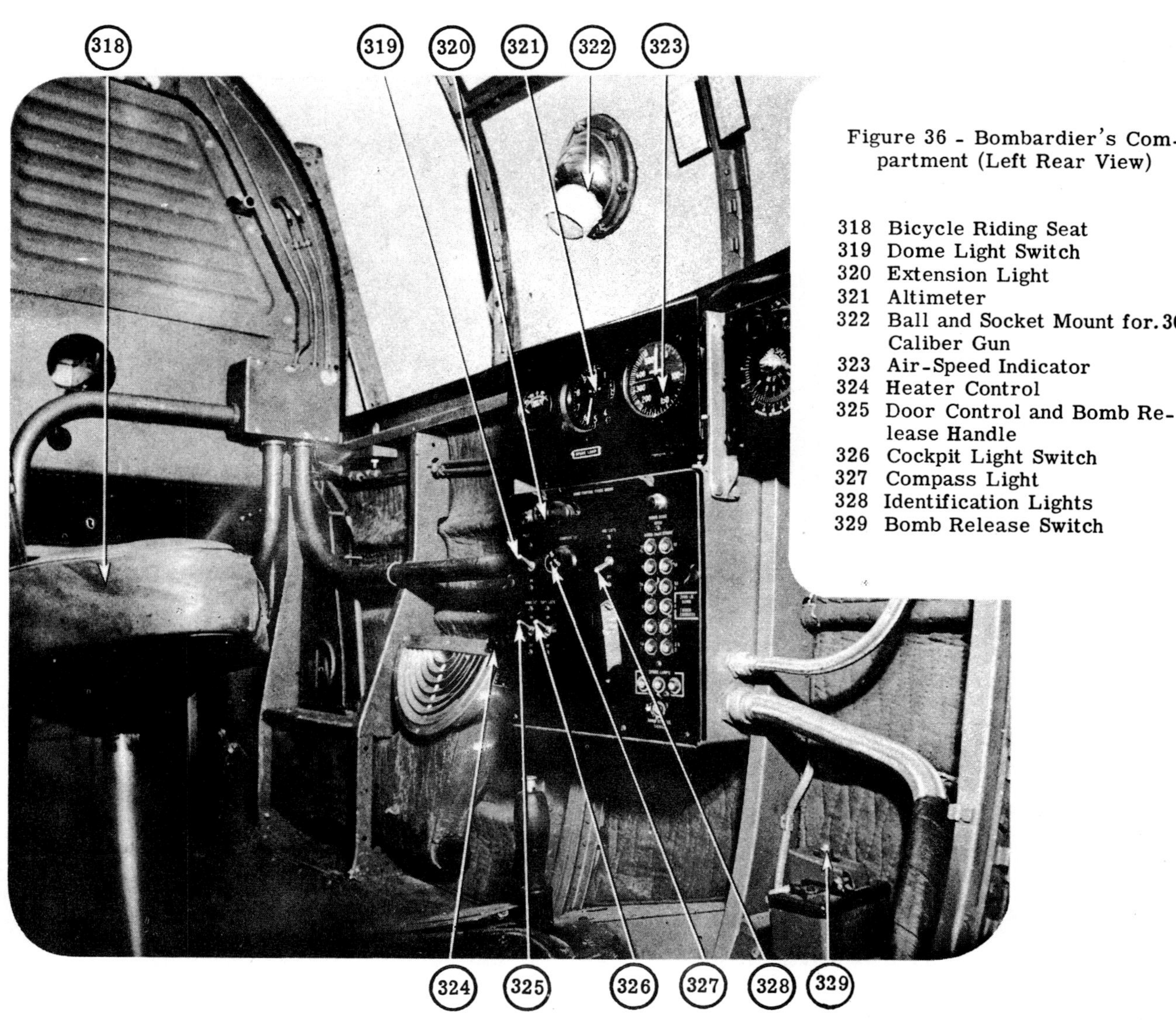

Figure 36 - Bombardier's Compartment (Left Rear View)

318 Bicycle Riding Seat
319 Dome Light Switch
320 Extension Light
321 Altimeter
322 Ball and Socket Mount for .30 Caliber Gun
323 Air-Speed Indicator
324 Heater Control
325 Door Control and Bomb Release Handle
326 Cockpit Light Switch
327 Compass Light
328 Identification Lights
329 Bomb Release Switch

Figure 37 - Bombardier's Compartment (Right Side)

330 Door for Cleaning Bomb Sight Window	335 Push-To-Talk Switch
331 Light Switch	336 Jack Box
332 Oxygen Regulator	337 .30 Caliber Machine Gun
333 Ammunition Box	338 Ball and Socket Mount for .30 Caliber Machine Gun
334 Oxygen Regulator	339 Bomb Sight Seat
	340 Riding Seat

SECTION VII

UPPER GUN TURRET COMPARTMENT

1. GENERAL DESCRIPTION.

a. The upper gun turret is located just forward of the armor plate bulkhead which separates the radio compartment from the photographer's and tail observer's compartment. It is equipped with oxygen and interphone controls.

b. This gun turret is an upper model "L" type and is provided with two (.50 caliber) model M-2 fixed machine guns equipped with hydraulically operated gun chargers and electric trigger motor solenoids. If the guns are elevated sufficiently to clear the fuselage, the turret may be rotated continuously (360 degrees) in azimuth. The guns swing from horizontal to straight up. The guns cannot be fired manually in the event of electrical failure. Provisions are made for 440 rounds of amunition per gun.

2. OPERATIONAL EQUIPMENT.

CAUTION: Do not operate turret without adequate power supply. Damage to the turret or airplane may result if the supply voltage at the turret is less than 20 volts.

a. Normal Operation of Turret.

(1) Before entering turret, turn on the disconnect switch (figure 40-349) located on the brush box at the forward side of the turret.

(2) Enter turret by pushing the riding seat to one side, and extend foot rests. (Do not step on the azimuth motor, compensator or brush box.)

CAUTION: Do not touch the main power switch on back of master control handle.

(3) Adjust foot rests and lock in place with the hand screw located at aft side of support attaching bracket. Adjust seat by sliding bracket on the center column.

(4) To charge the guns or remove faulty cartridge, depress gun charger button until chargers have moved gun bolts to rear position. Releasing the switch charges guns.

NOTE: Do not hold charger button depressed over 30 seconds.

(5) Rotate the sight cushion (figure 38) for individual focus.

b. Combat Operation.

(1) Move the master control handle (figure 39-346) to the neutral position and grip the control handle with the right hand to depress the main power switch on the back of the handle.

CAUTION: Do not depress the trigger switch on the front of control handle as the guns fire

as soon as the main power switch is depressed.

(2) The swing of the guns in elevation follows the swing of the control handle up or down, the speed being proportional to the degree of the handle movement from neutral. The guns swing from horizontal to straight up.

(3) To rotate the turret in azimuth, rotate the control handle about its vertical axis. The turret will turn in the same relative direction as the handle at a speed proportional to the degree of the control handle rotation from its neutral position.

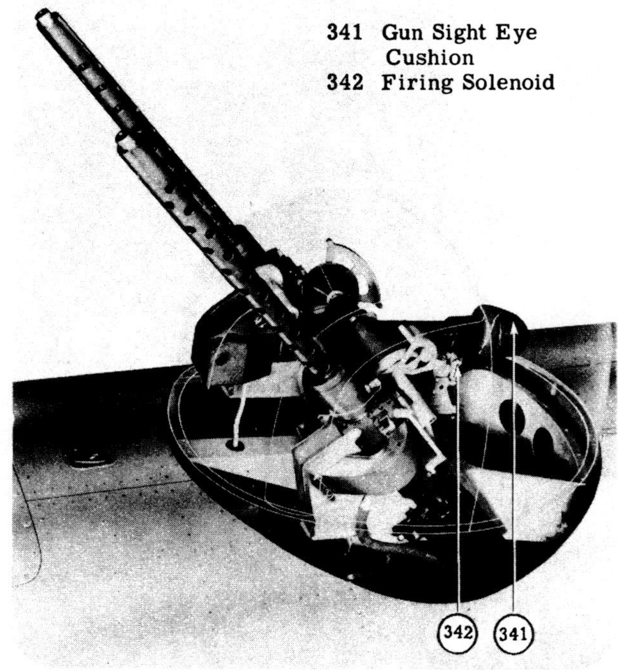

Figure 38 - Upper Gun Turret (Exterior View)

NOTE: The turret may be rotated continuously in azimuth (360 degrees) providing the guns are raised enough to clear the fuselage.

(4) To fire the guns, depress the trigger switch on the front control handle.

(5) Sight and train guns on the target by means of the master control handle. The intersection of the sight cross hairs indicates the point on which the guns are trained. (For sights without cross hairs use the dot in the center of reference circle.)

(6) Turn off the disconnect switch when the turret is not in use.

CAUTION: The electrical firing restrictor controls are disengaged when operating the turret mechanically. Make sure guns will not fire on airplane or strike fuselage. ▬ ▬ ▬ ▬ ➤

<u>c</u>. Emergency Operation.

NOTE: The crank stowed on the lower turret shall also be used for manual control of the upper turret. A crank extension shaft is provided and must be used for the upper gun turret control only.

(1) Engage the hand crank with the extension shaft in upper end of gun elevation motor shaft (figure 38-342) and rotate crank as required to position guns so that they will clear fuselage when turret is rotated to aft position.

(2) Engage the hand crank with the end of the shifter shaft and pull the shaft out one-half inch (1.27 cm) by means of the crank in order to disengage the shaft from the reduction gear mechanism. Rotate the turret so that guns are in their aft position.

<u>d</u>. Interphone Control. - The interphone jack box (figure 39-349) is located on the left wall of the airplane next to the turret. Operation is conventional.

<u>e</u>. Oxygen Control. - The oxygen regulator (figure 39-350) is next to the interphone jack box on the left wall. Operation is conventional.

☆   ☆   ☆

Figure 39 - Upper Gun Turret (Interior View)

345   Plug-In Cords
346   Master Control Handle
347   Parachute Stowage Space
348   Turret Operator's Seat
349   Jack Box
350   Oxygen Regulator

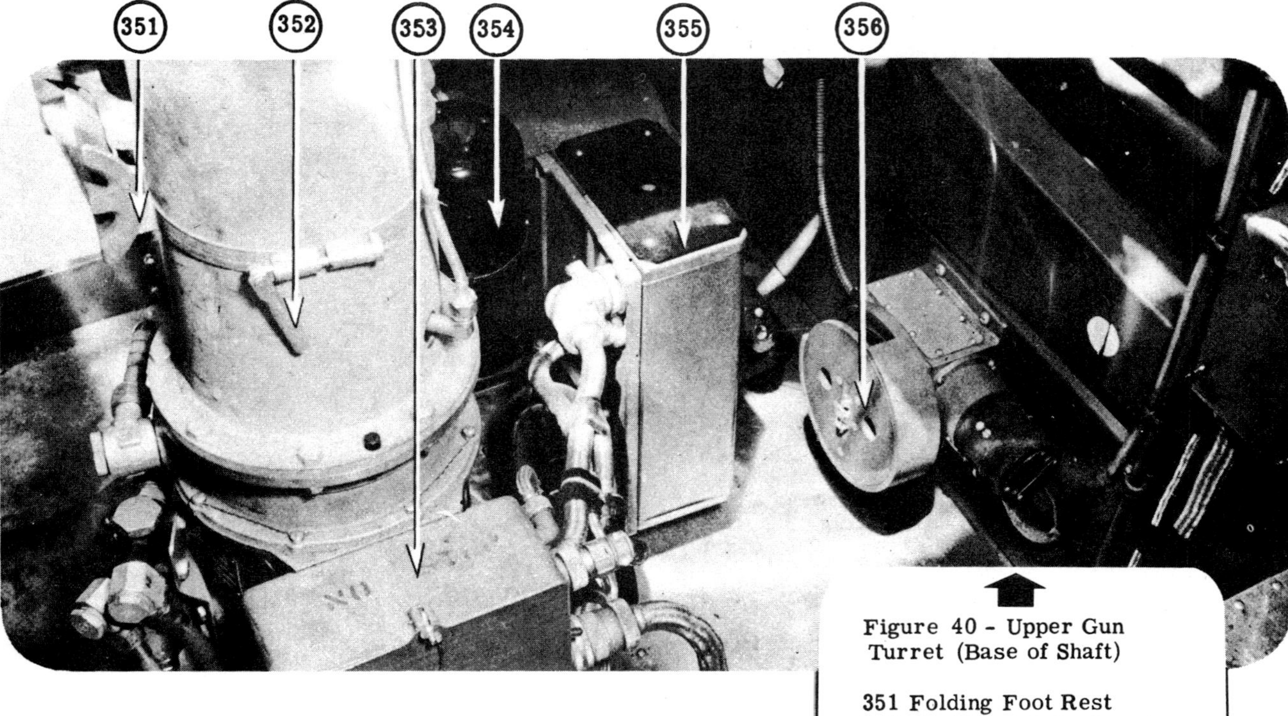

Figure 40 – Upper Gun Turret (Base of Shaft)

351 Folding Foot Rest
352 Foot Rest Adjustment
353 Brush Box
354 Turret Rotation and
    Retraction Motor
355 Compensator Box
356 Electrically Operated
    Trailing Reel Antenna

★ ★ ★

Figure 41 – Armor Plate Bulkhead

Armor Plate Door Immediately Aft of
the Upper Gun Turret

SECTION VIII

BOMB BAY

1. GENERAL DESCRIPTION.

a. The bomb bay is located between the navigator's and radio operator's compartments. The bomb bay may be used as a receptacle for bombs, for carrying the droppable fuel tank, or for the installation of the tow target equipment.

b. A passageway (figure 14) over the bomb bay is provided for the movement of the crew from station to station. A manhole in the passageway permits entering and leaving of bomb bay.

2. OPERATIONAL EQUIPMENT.

a. Bomb Bay Droppable Fuel Tank.

(1) The capacity of this tank is 585 U. S. gallons (488 Imperial gallons).

(2) This tank may be released in an emergency by either the pilot or bombardier.

b. Tow Target Equipment. - Emergency operation necessitates the following:

(1) Cut windlass cable.

(2) Escape through hatch in floor, then through bomb bay doors.

(3) If bomb bay doors are accidentally closed, exit is made through manhole in roof of bomb bay and then through one of the other hatches.

NOTE: An attachable parachute should be placed in the bomb bay whenever the tow target operator is at his station.

CAUTION: The bomb bay doors shall be open whenever tow target operator is in bomb bay, except during take-off or landing. Wire cutters must be carried by tow target operator at all times.

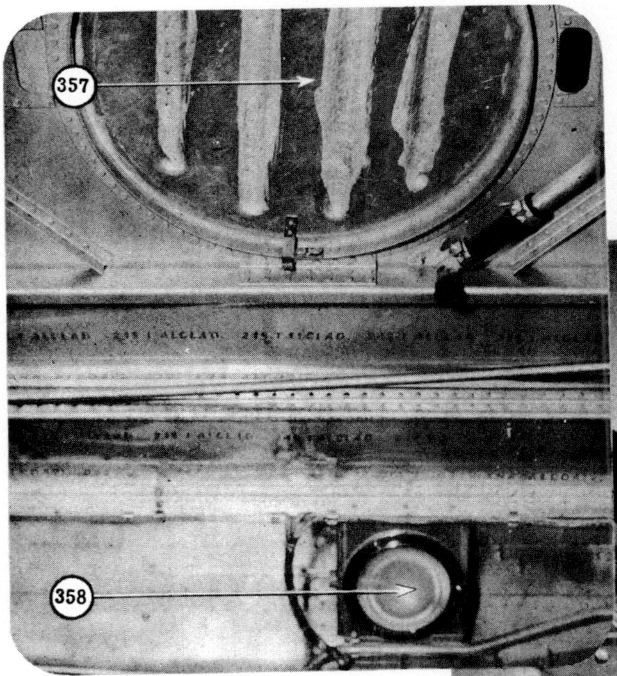

Figure 43 - Bomb Bay (Emergency Release for Bomb or Droppable Tank)

359 Dome Light Switch
360 Emergency Release for Droppable Tank or Bombs

Figure 42 - Bomb Bay(Hatch to Passageway over Bomb Bay)

357 Hatch to Passageway Over
     Bomb Bay
358 Dome Light

SECTION IX

PHOTOGRAPHER'S COMPARTMENT
(Including Tail Observer's Station)

1. GENERAL DESCRIPTION.

The photographer's compartment (figure 44) is located near the tail of the airplane just aft of the armor plated bulkhead at the rear of the upper gun turret, and forward of the tail observer's station. The tail observer's station is in the center of the empennage and the portion which extends beyond the elevators is enclosed with plexiglass. It affords a range of vision rearward of 180 degrees in azimuth and 180 degrees in elevation. No oxygen outlet is provided at the tail observer's station. A chemical toilet is located between the two compartments. One window on each side of the fuselage just aft of the camera equipment can be used for escape hatches on the ground only.

2. OPERATIONAL EQUIPMENT.

a. *Photographic Equipment.*

(1) Provision is made in the photographer's compartment for mounting a type K-7C, type K-3B or a type T-3A camera. A vacuum valve and camera power junction box (figure 44-363) in which are a signal light and an intervalometer (figure 44-360) and camera power sockets are provided. The camera is mounted on spring-loaded support tubes and may be raised to uncover the camera opening which has a removable cover. The cover can be stowed at the right side of the camera operator's seat. Windows are provided on each side of the fuselage for taking oblique photographs. These windows hinge inboard from the top and are held open by hooks.

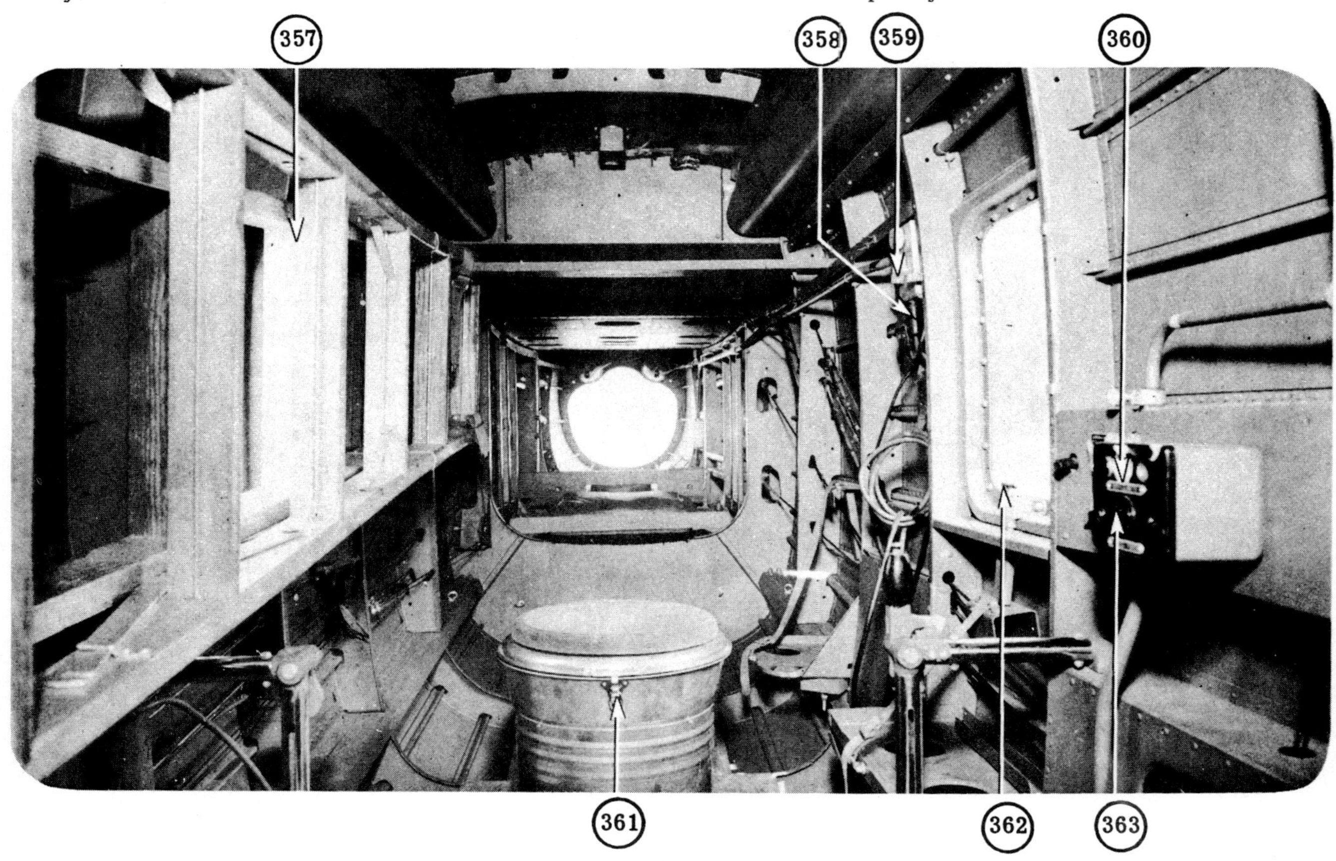

Figure 44 - Photographer's Compartment (Rear View)

357 Service Stepladder	360 Intervalometer
358 Push-To-Talk Switch	361 Chemical Toilet
359 Jack Box	362 Side Window Latch
363 Camera Power Junction Box	

CAUTION: When windows are closed, stow
the hooks to prevent injury to crew members.

    (2) A camera rest is secured to either window
sill with fasteners.

    (3) The chemical toilet (figure 44-361) is used
as the camera operator's seat. No safety belt is provided for this seat.

    b. Interphone Control. - A jack box (figure 44-359)
is located on the left wall aft of the escape hatch.
Operation is conventional.

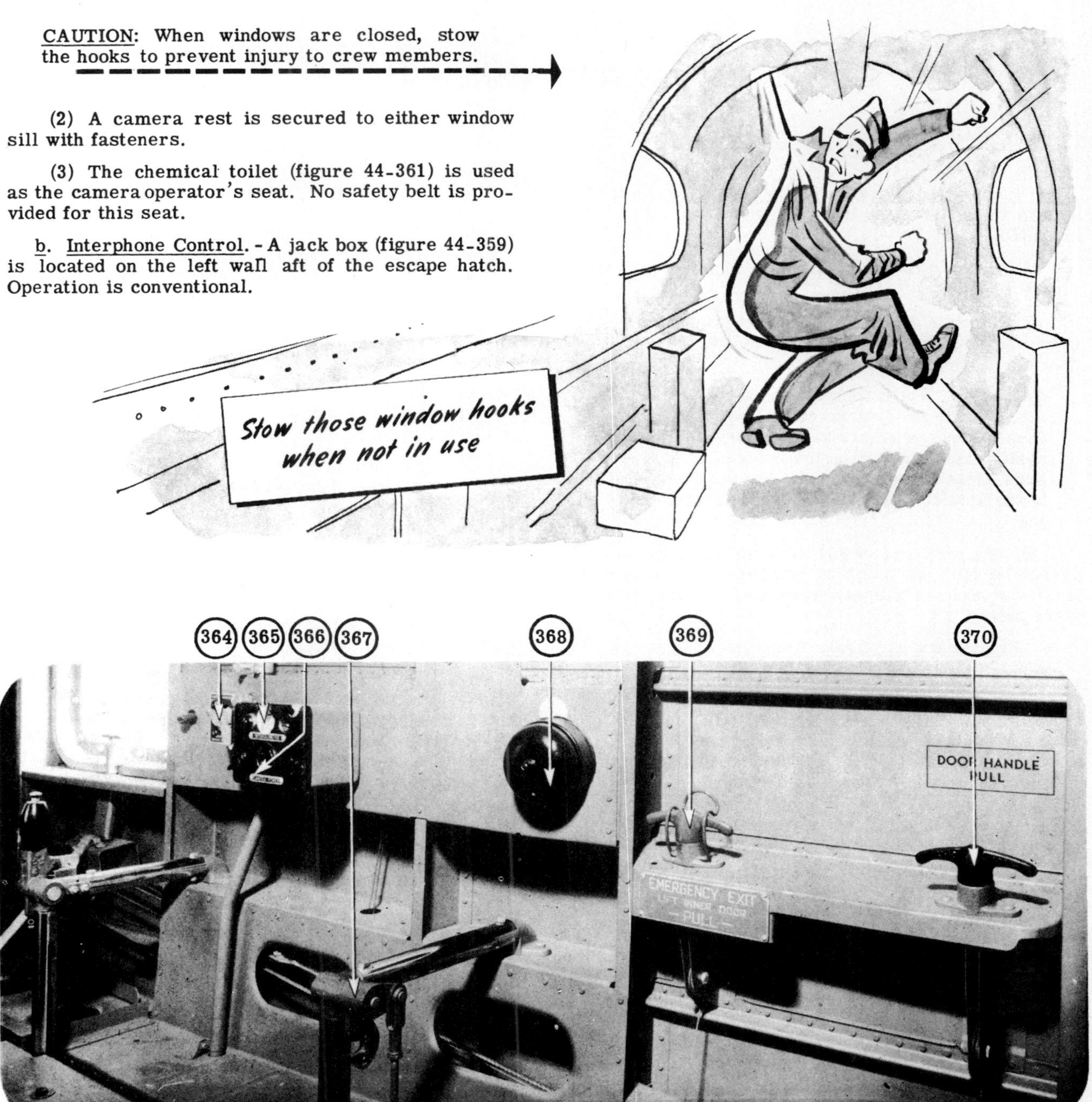

Figure 45 - Photographer's Compartment (Right Side)

364	Camera Vacuum Selector Switch	367	Camera Mount
365	Intervalometer	368	Alarm Bell
366	Camera Power	369	Emergency Exit Door Pull
		370	Door Pull Handle

SECTION X

EMERGENCY OPERATIONS AND INSTRUCTIONS

1. EMERGENCY OPERATION IN CASE OF HYDRAULIC SYSTEM FAILURE

a. **Emergency Hydraulic Brake Operation.** - If there is less than 800 lb/sq in. (56.2 kg/sq cm) pressure indicated.

(1) Turn emergency hydraulic selector valve (figure 21-200) to "BRAKE."

(2) Operate hydraulic hand-pump until pressure gages show 800 lb/sq in. (56.2 kg/sq cm) and not more than 1450 lb/sq in. (102 kg/sq cm).

CAUTION: While applying brakes after landing, operate hand-pump continuously to maintain as much pressure as possible in the brake system accumulator as the accumulator pressure alone is not adequate for the amount of brake application required for a normal landing.

(3) If a pressure of 600 lb/sq in. (42.2 kg/sq cm) cannot be built up prior to landing, a field with at least one mile of runway is needed to land the airplane safely.

b. **Emergency Air Brake Control.**

CAUTION: Use air pressure brake system as a last resort only. When it is known that the emergency air pressure brake system may have to be used during landing, the longest runway available shall be sought and the shortest landing possible made. As brakes cannot be applied selectively, the pilot must be ready to counteract any uneven action with the throttle.

(1) **To Apply Emergency Air Brake.** - Pull up sharply on control handle (figure 21-193), breaking safety wire, and lower handle halfway. Repeat this operation, applying brakes by very quick, successive upward pulls of the handle from the intermediate position until the desired amount of brake action is obtained. Spring action aids in lowering the handle to halfway (intermediate) position, in which position the air pressure to the brakes will be maintained.

(2) **To Release the Emergency Air Brakes.** - It is only necessary to push handle fully down to its normal position.

(3) The air brake may be reapplied but the pilot should depend on only one application.

CAUTION: After a landing during which the air brakes were used, taxying should be accomplished with extreme care, as very little, if any, brake pressure will be available.

CAUTION: After using emergency air brake system, hydraulic brake system must be bled.

c. **Main Landing Gear Emergency Mechanical Operation.**

NOTE: The screw jack for lowering the main gear is very powerful, and if the handle continues to be turned after the gear is down and locked, damage to the cable results. Contact between the pilot and the radio operator by the interphone system is imperative. It is also important not to return the lowering screw handle to its original position until the airplane is safely landed.

(1) Move hydraulic landing gear control handle (figure 21-192) in pilot's compartment to "DOWN" position.

(2) Pull pin and fold down adjustment rod of lower turret operator's chest support.

(3) Lower radio operator's table to useful position.

(4) Release main landing gear operating screw assembly (figure 29-267) (held to forward wall of radio compartment with finger type fastener) and rotate assembly away from wall until it locks in a fore and aft position.

(5) Pull main landing gear up-position latch release, located adjacent to lowering screw. This control releases the main gear up-position latches only, and allows the main gear to partially lower due to its own weight. Check visually or on pilot's landing gear position indicator that both main gears are partially lowered before using the lowering mechanism.

(6) Turn lowering screw lever clockwise (figure 47) until indicators register the main gear is down and locked.

CAUTION: Do not lower main landing gear by the emergency mechanical lowering system above 150mph (130 knots) indicated air speed.

NOTE: The main gear cannot be raised mechanically.

d. **Nose Wheel Emergency Mechanical Operation.**

(1) Ascertain that the landing gear control handle (figure 21-192) in pilot's compartment is in the "DOWN" position.

(2) Pull nose gear emergency lock release (figure 25-238) located on top of mechanism. This releases the nose gear and allows it to lower partially due to its own weight. Check on pilot's gear indicator or by means of the drift meter, to see that auxiliary gear is partially lowered.

(3) Remove safety pin and turn pawl (figure 25-239) to "ON."

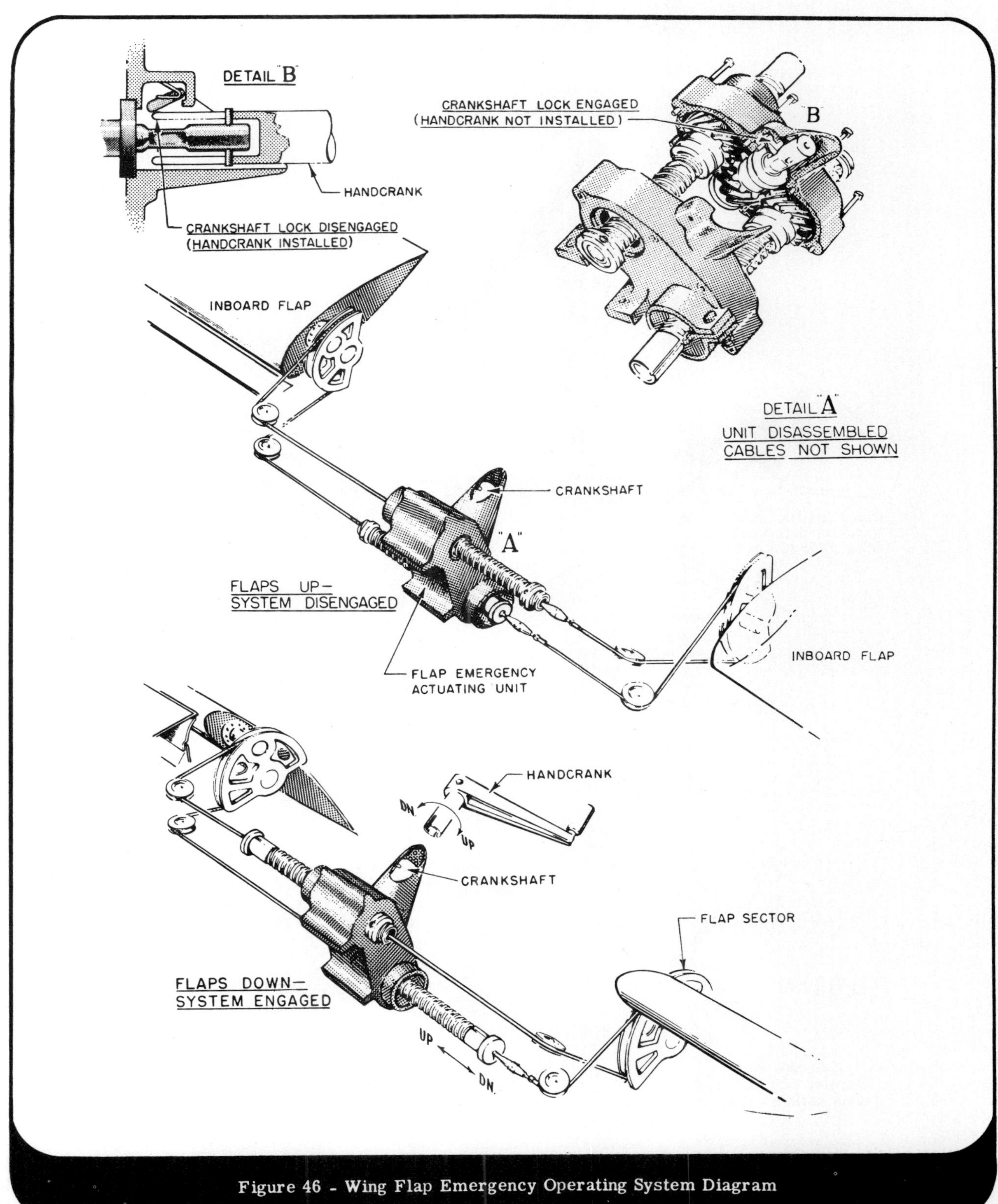

Figure 46 - Wing Flap Emergency Operating System Diagram

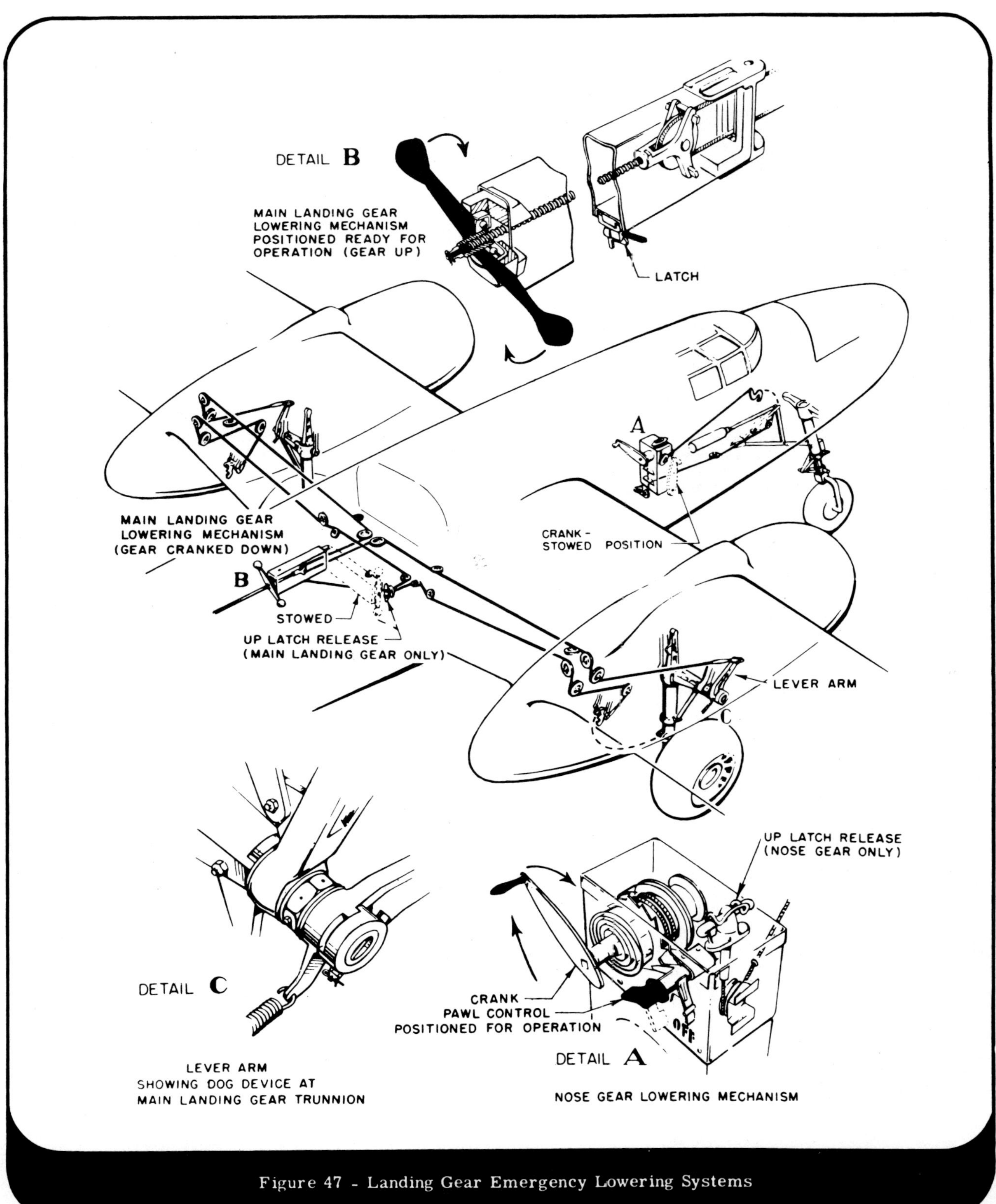

Figure 47 - Landing Gear Emergency Lowering Systems

(4) Place crank (figure 25-241) on shaft and turn clockwise until gear indicator registers "DOWN" and "LOCKED."

For emergency nose wheel operation:-
Turn handle clockwise

CAUTION: Do not lower nose gear by means of the mechanical lowering system, above an indicated air speed of 150 mph (130 knots). Do not return pawl to "OFF" until airplane is safely on the ground.

e. **Emergency Hydraulic Operation of Main Landing Gear Down-Position Latches.**

NOTE: Whenever the main landing gear down-position lockpins fail to engage automatically, regardless of the method used to lower the main gear, the following operations should be performed.

(1) With gear fully extended (check position indicator) turn emergency hydraulic selector valve (figure 21-200) to "LATCH."

(2) Operate hydraulic hand-pump until position indicator registers main landing gear lockpins are in place.

NOTE: Main landing gear must be fully down, prior to using hand-pump.

CAUTION: Once the lockpins have been pumped into the latched position, they cannot be retracted. It is impossible to get the gear into the fully down position if the latches have been pumped to the latched position prematurely.

NOTE: If the nose wheel position indicator shows the nose landing gear fully extended but warning flag shows the down-position lock is not engaged, it is safe to land if the pawl of the nose wheel emergency lowering crank is "ON" and the gear has been pulled down as far as possible. Before landing under these conditions, send all available personnel into the tail section of the airplane.

f. **After Landing.** - Before operating the landing gear hydraulically, again it is absolutely essential to perform the following:

(1) Turn nose gear emergency lowering mechanism pawl control to "OFF" position and stow crank on side of drum housing. To turn pawl control "OFF" relieve tension on pawl by exerting pressure on crank handle.

(2) Turn main landing gear emergency lowering screw handle counterclockwise to return handle to its original position and stow assembly with fastener provided.

g. **Bomb Bay Door Emergency Mechanical Operation.** - To be used only in the event of complete hydraulic pressure failure.

(1) Move bomb release handle to the "DOORS OPEN-RACK LOCK" or the "DOORS CLOSED" position as desired.

(2) Install bomb bay door operating crank on shaft in aft end of navigator's compartment. (Crank is stowed on lower right longeron opposite the crankshaft.) The crank shall be installed with the crank positioned downward. Turn crank clockwise to close doors and counterclockwise to open doors. The crank has an automatic clutch mechanism which enables it to be operated in either direction and helps to hold the doors in the position to which they are operated.

NOTE: After closing the bomb bay doors by means of the crank, thread strap secured to crank handle through tie-down loop on floor below crank and tighten securely.

h. **Wing Flap Emergency Mechanical Operation.**

CAUTION: Do not attempt to operate the emergency flap operating system (do not even install crank) unless there is hydraulic system pressure failure, as the hydraulic and mechanical systems oppose each other.

(1) Coordination through interphone communication between pilot and radio compartment is necessary.

(2) Reduce speed of airplane to 150 mph (130 knots) before operating flaps by means of emergency system. Do not fly at a speed over 150 mph after flaps are mechanically lowered.

(3) Move pilot's hydraulic flap control handle to the extreme "DOWN" position.

(4) Open hinged cover located on shelf at forward

end of radio operator's compartment, and install crank (stowed on wall below hinged cover) on shaft of flap operating mechanism.

(5) Rotate the crank clockwise to lower flaps. (Approximately 27 full turns of the crank are required to lower flaps.)

(6) To lock the flaps in any desired position, remove the hand crank.

(7) To disengage the mechanical flap operating mechanism, it is necessary to rotate the crank counterclockwise against its stop and stow the crank. Inadvertent engagement of the mechanical system is prevented by a lock which acts upon the removal of the hand crank. This lock also serves as guarantee against engagement of the mechanical system while the hydraulic system is in operation, if the crank has been turned counterclockwise as far as it will go.

CAUTION: The emergency mechanical flap operating system must be returned to its full "UP" position, by turning the crank counterclockwise against its stop, after which the crank must be removed and stowed before the wing flaps are operated hydraulically again. Otherwise damage to the cable system will result.

(8) Prior to lowering the wing flaps mechanically, it is necessary to be sure that the pilot's hydraulic flap control handle is moved to the extreme "DOWN" position. Approximately one minute in flight is required to lower the flaps mechanically.

i. **Hydraulic Operation of Wing Flaps with Engine Stopped.**

(1) Move wing flap control handle (figure 21-191) to the extreme position for the desired movement. Normally, there will be enough pressure in the accumulator (800 to 1100 lb/sq in., 56.2 to 77.3 kg/sq cm) to complete one full movement of the flaps. Further movement of the flaps should be accomplished as follows.

(2) Turn emergency hydraulic selector valve to "NORMAL."

(3) Operate hydraulic hand-pump until the desired position of the flap is reached, after which flap control handle shall be returned to "NEUTRAL."

NOTE: To lower wing flaps, by the emergency mechanical lowering system, while on the ground all pressure must be bled from the hydraulic system. This may be done by operating the engine cowl flaps until gage registers zero.

2. **EMERGENCY EXIT DURING FLIGHT.**

a. Pilot, copilot, bombardier and navigator use front entrance hatch located in the floor of navigator's compartment.

(1) Raise inner door by pulling lever (figure 22-210) and secure door with catch.

(2) Disengage safety guard on emergency release handle (figure 28-252) and pull handle.

b. All crew members stationed aft of the navigator's compartment should use rear entrance hatch aft of upper turret bulkhead.

(1) Raise inner door by pulling lever (figure 45-370) and secure door with catch.

(2) Disengage safety guard on emergency release handle (figure 45-369) and pull handle.

(3) The escape hatch located at the right side of the lower turret releases inboard.

c. The following procedure is to be carried out by the tow target operator.

(1) Cut windlass cable if trailing.

NOTE: Wire cutters must be carried by tow target operator at all times. Bomb bay doors must be kept open while operator is in bomb bay, except during landing or take-off. If possible, cut or retract trailing antenna before using floor hatches.

(2) Escape through hatch in floor aft of seat, then through bomb bay doors.

(3) If bomb bay doors are accidentally closed, make exit through manhole in roof of bomb bay and then through one of the other hatches.

CAUTION: To use bombardier's side escape hatch, pilot's escape hatch, or pilot's side windows as a means of escape during flight, propellers must be feathered and the airplane must be under control.

3. **EMERGENCY EXIT ON GROUND.**

In addition to the above exits of escape are the following ground exits:

a. Bombardier's enclosure escape hatch located on the left side of the compartment. This hatch releases inboard.

b. Pilot's enclosure escape hatch is located above pilot's and copilot's seats. The hatch releases outboard.

c. Pilot's compartment side windows slide aft on tracks. To operate, squeeze handles located forward of windows near the instrument panel.

d. The side windows in the photographer's compartment which hinge at the top may be opened from within the airplane and used as exits.

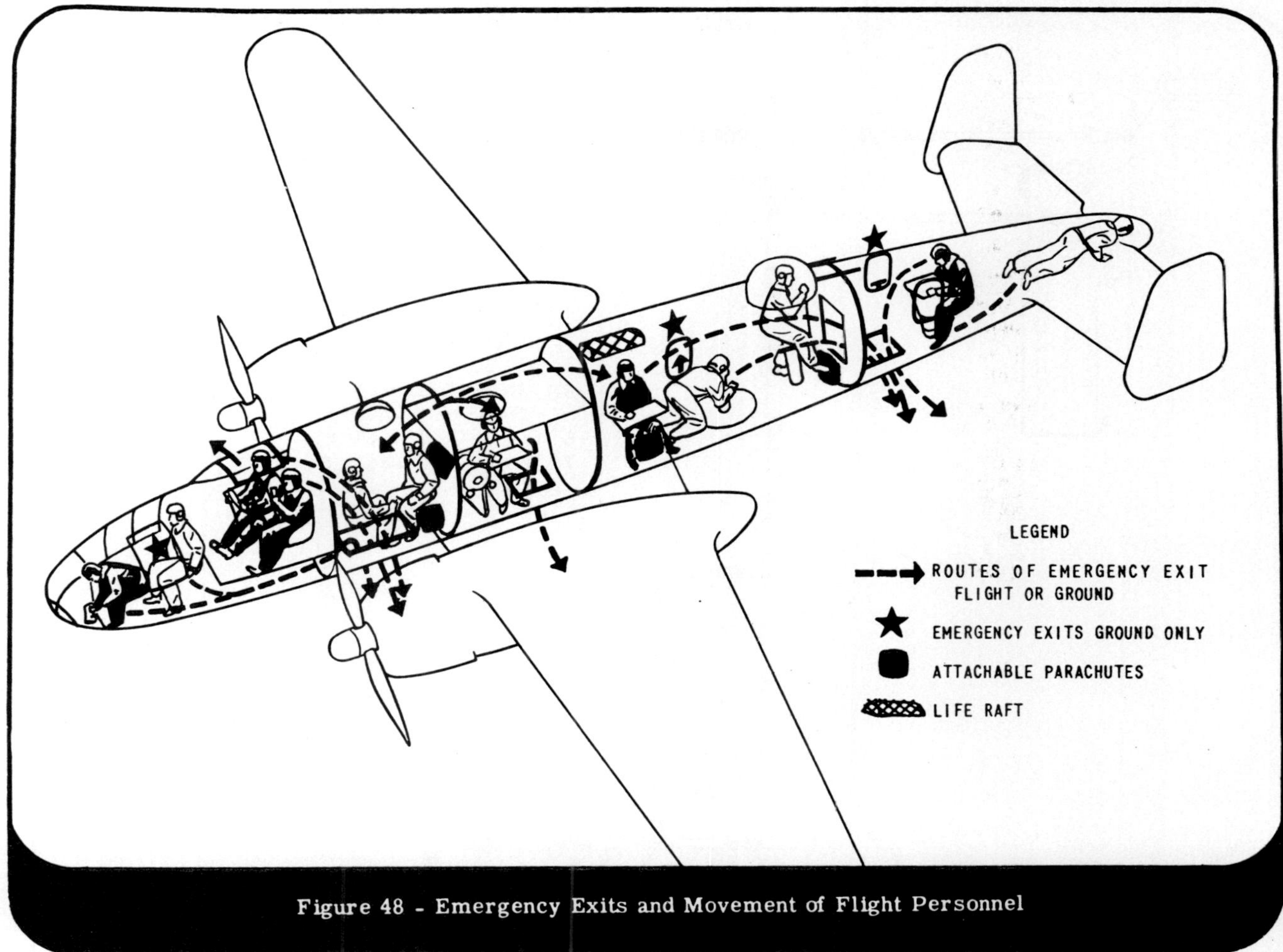

Figure 48 - Emergency Exits and Movement of Flight Personnel

4. MISCELLANEOUS EMERGENCY NOTES.

a. Interphone.

(1) Set jack box dial control to "CALL."

(2) All stations, even those already in use, will be placed in parallel across the amplifier.

b. Emergency Fuel Shut-off Valves.

(1) Valves (figure 23-227) are located in right-hand rear corner of navigator's compartment.

(2) To operate: depress spring-loaded catch (figure 23-228) and twist control handles to "OFF."

c. Emergency Release of Bombs or Droppable Fuel Tank.

(1) All bombs may be released simultaneously and in "SAFE" condition by operating pilot's emergency release (figure 16-102) or bombardier's release control. (See figure 35-314.)

(2) To operate bombardier's release control.

(a) Hinge the anti-salvo guard upward and move the bomb release handle to the extreme forward "SALVO" position.

(b) In case of hydraulic system failure, crank bomb bay doors open mechanically by the following procedure:

1. Ascertain that bombardier's door-operating and bomb release handle is at the "DOORS OPEN-RACK LOCK" or "DOORS CLOSED" position.

2. Install bomb bay door operating crank on shaft (figure 22-204) in aft end of navigator's compartment. Crank is stowed on lower right longeron opposite the crank shaft. The crank shall be installed with the crank positioned downward. Turn crank clockwise to close doors and counterclockwise to open doors. The crank incorporates an automatic clutch mechanism which enables it to be operated in either direction and helps to hold the doors in the desired position.

WARNING: After closing the bomb bay doors

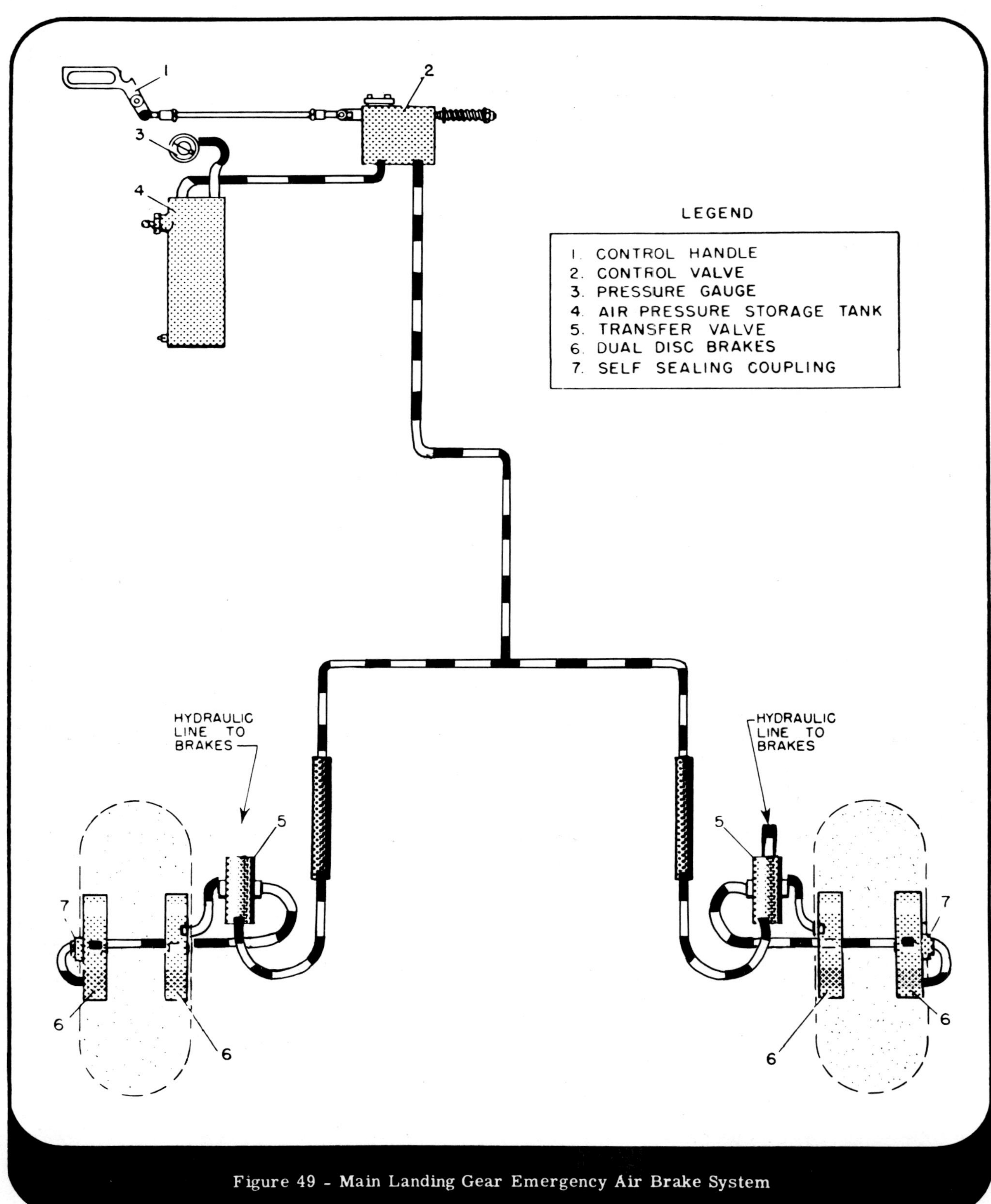

Figure 49 - Main Landing Gear Emergency Air Brake System

by means of the crank, thread the strap secured to crank handle through tie-down loop on floor below crank and tighten securely. This is necessary as the brake action of the crank clutch mechanism is insufficient to hold doors closed.

(3) If emergency bomb release has been sprung, or if the bombardier's control handle has been placed in "SALVO" the bombs or bomb bay droppable tank will release automatically when doors reach full open position.

NOTE: If the automatic release is not desired, before cranking doors open, recock pilot's emergency release by moving bomb release control forward past the "SALVO" position and then back to "DOORS OPEN" position.

d. Alarm Bells. - Three emergency warning bells are located, one each in the bombardier's radio operator's, and photographer's compartment. The control switch is located on the pilot's switch panel.

e. Fire Extinguishers.

(1) The engine fire extinguisher control (figure 16-144) is located below the instrument panel forward of the copilot. To operate, set valve to "RIGHT ENGINE" or "LEFT ENGINE" and then pull handle.

(2) A carbon dioxide fire extinguisher is located at the right side of the navigator's compartment and on the right side of the radio operator's compartment. To remove a fire extinguisher from its bracket, push up on retainer strap handle. To operate:

(a) Swing horn up.

(b) Discharge close to base of fire.

(c) Control with trigger.

Watch how you handle a CO₂ extinguisher. Look out for frostbite!

CAUTION: White discharge is dry ice. Avoid frost bite. Do not permit extended contact with skin.

(3) One carbon tetrachloride fire extinguisher is mounted on the inner side of the "RED" door hinged on the lower outboard side of each engine nacelle.

WARNING: Do not use carbon tetrachloride fire extinguisher in an enclosed space since the gas formed by this chemical is phosgene which acts as an anesthetic and may result in fatal poisoning.

f. Life Raft Release. - A type A-2 life raft is stowed in the upper left side of the fuselage in the radio operator's compartment.

(1) To operate from within airplane pull control (figure 30-273) in radio operator's compartment.

(2) To operate when outside airplane pull control located on left upper side of fuselage above the trailing edge of wing.

5. FORCED DESCENT OF LAND PLANES AT SEA (DITCHING).

a. General. - The following notes have been prepared by the British for the general guidance of all members of airplane crews in the event of a forced landing at sea, which is called "Ditching". These general notes do not specifically apply to this type of airplane. The life raft referred to in paragraph 4.f. is called a "Dinghy" in the following text. American airplanes at present have no provision for dumping fuel except by releasing the bomb bay droppable tank.

b. Preparation for Ditching.

(1) If doubt exists in the pilot's mind whether he can reach the coast, preparation for ditching must begin, particularly as to radio procedure.

(2) If height cannot be maintained above 1000 feet the crew should move to their stations in order that the pilot can readjust trim, and lower his flaps without the crew moving about the airplane.

(3) The pilot's command to prepare for ditching is "Dinghy, dinghy, prepare for ditching" which must only be given by the pilot. The command will be acknowledged by the entire crew on the interphone system with the answer "Navigator ditching" or "Bombardier ditching," whichever is appropriate. The crew should also have a prearranged call light ditching signal and the letter "D" repeated three times is appropriate. The pilot will normally warn the wireless operator in this manner and the member of the crew nearest the wireless operator should also give him verbal warning. The preparation for ditching is thus begun on a coordinated basis and the pilot is assured that his crew are aware of the situation, and if they have practiced the drill, that they know what to do and do it.

(4) The pilot's duty is to coordinate the work of his crew, but each crew member should act on the pilot's command "Dinghy, dinghy, prepare for ditching"

without further orders being necessary, other than the pilot's final command to the wireless operator to move to his ditching station and the final warning of the impending impact.

   c. The Navigator.

   (1) The navigator should have a constant knowledge of the wind speed, direction, drift and the fixed position of the airplane. He should always know the fuel consumption in relation to his estimated time of arrival.

   (2) At the pilot's command the navigator will:

   (a) Calculate his position.

   (b) Advise the wireless operator of the drift position, with the course, height, and speed maintained.

   (c) Receive fixes and bearings from wireless operator.

   (d) Calculate estimated position of ditching and advise wireless operator.

   (e) Inform pilot of surface wind speed and direction.

   (f) Make out air and dinghy release pigeon messages.

   (g) Destroy secret papers and place charts (with latest position marked thereon) in satchel.

   d. Wireless Operator. - On the pilot's command "Dinghy, dinghy, prepare for ditching" the wireless operator will:

   (1) If on "GROUP" frequency make the first signal on that frequency and then change over to the allotted MFDF section.

   (2) Turn IFF to emergency.

   (3) According to the situation use one of the three priority calls:

   (a) S.O.S. I am in immediate need of assistance. May Day (by radio telephone).

   (b) I may require assistance.

   (c) I may be forced to land without further signal.

   (4) Give a time and position to the signal. It is better to make one of the appropriate distress signals, than to remain silent. A distress call can always be cancelled when no longer applicable and in fact, this must be done.

   (a) Transmit course, height and ground speed maintained.

   (b) Advise navigator of fixes or bearings.

   (c) Get estimated position of ditching from navigator.

   (d) Transmit estimated position of ditching.

   (e) Clamp down key on pilot's command and move to ditching station.

   (f) Destroy secret papers.

   (g) Where possible use the trailing antenna as an altimeter.

   e. It is the Personal Responsibility of the Pilot:

   (1) To be sure that the bomb doors are opened, the bombs and containers jettisoned and the doors closed again. It takes time to open and close doors and if there is any doubt that there is sufficient time to accomplish this, it is better to keep the doors closed; in this case it is essential for the pilot to check the bomb controls at "SAFE."

   (2) To determine whether or not to jettison fuel, the crew member who has been detailed in the previous drill opens the cocks on the pilot's order. After the fuel is jettisoned it is imperative that the cocks be closed again to retain the buoyancy of the tanks. Fuel cocks take time to open and close and fuel can seldom be jettisoned faster than 100 gallons per minute. If the airplane is equipped with a full droppable tank, drop it, but if empty retain to aid flotation.

   (3) Make sure that the crew member detailed in the drill assists him to secure his sutton harness.

   (4) To release pilot's upper escape hatch.

   (5) Check that landing gear and nose gear is "UP."

   (6) Lower flaps as required.

   (7) To order the wireless operator to his ditching station, since it is important that he remain at the set as long as possible.

   (8) To warn the crew when ditching is imminent.

   (9) To switch on the landing lights and upper identification lights (if this doesn't cause reflections which upset vision). It is important to remember that although the surface may be seen in the beam of the landing lights, judgment of height may not be correct.

   f. Preparation of the Airplane to Make it as Seaworthy as Possible.

   (1) Not only does jettisoning the fuel lighten the airplane and so reduce the speed at which the airplane may be ditched, but also the empty fuel tanks are a considerable contribution to flotation.

   (2) The security of all lower and side hatches must be checked. Side escape hatches may have to be used in ditching but only upper escape hatches can be regarded as ideal, since they must be opened before ditching. This is necessary because the hatches may become jammed on impact and also because it is essential for the crew to be free to leave the airplane without delay after ditching.

   (3) The bombs should be jettisoned to lighten the airplane to assist in reducing the air speed at impact and this loss of extra weight will contribute considerably to flotation. If there is any danger of the bomb doors being open when the airplane hits the water, it is better to keep the bombs on board in "SAFE" position. Thirty seconds must be allowed for the opening

and closing of the bomb doors. All loose equipment should also be jettisoned for similar reasons.

NOTE: Be sure that when equipment is jettisoned it does not hit the empennage or carry away the IFF antenna.

(4) All bulkhead doors must be closed to hinder the flow of water, from bow to stern.

(5) Close all camera hatches and flare chutes.

g. <u>Preparation by the Crew to Insure Safety On and After Impact.</u>

(1) All the actions to make the airplane seaworthy also come into this category.

(2) It is vitally important that the crew should be braced against the impact. There are two ideal ditching stations:

(a) In a sitting position back and head braced against a solid structure such as at the rear of a spar, an armored door or an armored bulkhead. If the head comes above a spar being used as a ditching station, it is very important that the head should be clasped in the hands to avoid it being forced back and injured. In this position the body can withstand forces which are far greater than those expected in ditching with the exception of forces expected when the airplane dives straight in.

(b) The second but less satisfactory ditching station is to lie on the floor with the head to the rear and the feet braced against a solid structure. It is necessary to have the knees bent to avoid injury as far as possible, but the limiting factor of this ditching station is the liability of the legs to fracture.

(3) Straps are not normally required at ditching stations unless there is a lack of suitable positions in the airplane in which case the crew member may have to remain in his seat. Loss of life may occur due to failure to get clear of the airplane so that straps must not be used unless virtually necessary.

(4) It is vitally necessary that the pilot be secured by sutton harness and it is considered that the embarrassment caused by having harness done up during ditching is far less serious than the consequence of not being secured.

(5) The rear step formed by the end of the bomb bay should not be used as a ditching station, since a great rush of water is expected here, owing to the almost certain collapse of the bomb bay doors and consequently the step will be liable to burst inward.

(6) All forward and amidship upper hatches should be opened before ditching to facilitate the rapid egress of the crew and also to insure that the hatches do not become jammed on impact due to being left closed. It should, however, be borne in mind that open hatches cause drag and therefore, if the airplane is being flown at reduced power these upper hatches should not be opened until at least 1000 feet is reached.

(7) In night ditching, all bright internal lights should be put out and only the amber lights used. This will accustom the eyes to the external darkness.

(8) All lights should be left on after ditching to facilitate search, in the event of the airplane floating for a period.

(9) Life jackets must be worn at all times with the leg straps secured. Where there are small upper ditching hatches, jackets should not be inflated until immediately after leaving the hatch. On airplanes with large upper hatches the jacket may be inflated before the ditching takes place. In most cases it is safe to inflate the jacket with one or two breaths before ditching.

(10) Parachute harnesses should be removed before ditching in all cases where practicable, except where the single seater dinghy is attached to the parachute harness.

(11) Helmets should be retained for the sake of protection of the head against cold when in the dinghy. The leads should be tucked firmly within the life jacket below the V of the neck, at the top tie.

(12) The latest airplane sea rescue equipment is usually stowed in either the dinghy stowage or conveniently near the ditching hatches and it should not be removed from these stowages before ditching to avoid it being flung forward on impact and becoming lost in the surge of water. That equipment which is carried free must be held firmly during ditching.

h. <u>Wind Speed and Direction and Surface Conditions in Relation to Ditching.</u>

(1) At least an elementary understanding of sea conditions must be gained to obtain full advantage from the notes on handling, which follow this section.

(2) With a calm type of sea, there may be little or no wind, so that it is essential to ditch with the lowest indicated air speed possible. Such a sea is deceptive with regard to judgment of height, particularly if the surface is glassy. If there are ripples upon the surface judgment of height is improved.

(3) Waves always move with the wind except close inshore and in fast flowing estuaries. Waves are the direct result of the wind which creates them and maintains them.

(4) "Swell" is an undulating movement of the surface caused by past or distant disturbance by action of the wind. It does not necessarily move with the wind and it has no breaking crests. If the wind is blowing across the swell a cross-sea is created with the waves (which are moving downwind) running on the swell. In these conditions the pilot must choose that direction along the swell which will make the approach as near into the wind as possible.

i. <u>Ditching Characteristics.</u> - If the airplane alights tail down in a three pointer attitude (as it should) there will be a primary slight impact as the rear of the airplane strikes. This will be followed by a severe impact with violent deceleration in most cases. If the alighting has been made too fast a bounce will occur, providing the fuselage is sufficiently strong. As the airplane comes to rest the nose will bury, but if the alighting has been carried out correctly, the effect of the nose burying will be minimized and the structure may not collapse. Bomber airplanes may usually be expected to float for a minimum period of one minute.

j. **Characteristics in a Short, Moderate or Calm Sea.** - If the airplane bounces, the control column should be held back. In the average short sea the tail should touch the crest of a wave and as soon as it does the nose should be kept up as much as possible. This should cause the forebody to touch down approximately under the center of gravity on the next wave.

The sea's not as calm as it appears from up there

WARNING: The open sea always appears from the air to be much more calm than is the case.

k. **Wind, Speed, and Direction.**

(1) In the absence of any fixed mark (land, lightship, etc.) or floating object not under way, the pilot can only judge motion relative to the motion of the waves.

(2) Waves, as distinct from swell, move downwind and the line of the wind can be taken to be at right angles to the lines of the wave crests; but doubt may exist as to which way the wind blows along the line.

(3) If there is sufficient wind, waves break and they break downwind. This can readily be observed from a low height. If the airplane is flown at right angles to the breaking waves the direction of drift will be apparent.

(4) If there is enough wind to blow the spray off the wave crests, the direction in which the spray moves is reliable.

(5) Wind direction may be obtained by dropping a smoke float. The smoke from ships is also a useful guide. Smoke naturally drifts with the wind and if this drift could be observed the direction would be indicated. But do not make the mistake of supposing that the wind direction is along the trail of the smoke. This trail is resultant of the wind speed and direction and the ships forward motion. Therefore the wind direction is somewhere between the forward path of the ship and the smoke trail. Only when the wind is blowing in a similar direction to the forward motion of the ship will the smoke be a reliable indication of direction. It will be from astern.

(6) If low enough it is possible to calculate the direction of the wind by observing the sails of surface craft. A reasonable indication of speed can also be gained by observing the set of the sails.

(7) Where the surface is not broken up, it is possible to watch gusts rippling the surface in great sweeps, which indicate the wind direction.

l. **Drill During the Final Approach.**

(1) The pilot should keep his wireless operator at the set as long as possible and only allow him a safe margin of time to take up his ditching station.

(2) The crew on their part must see to it that the wireless operator's ditching station is not occupied and is clear of obstacles.

(3) The pilot will warn the wireless operator to move to his ditching station by call light and / or interphone; or by shouting.

(4) The wireless operator for his part can be fairly certain that the order will come when he feels the flaps finally being lowered.

(5) The wireless operator will immediately clamp down the key and move to the ditching station at the pilot's command, fully realizing that he has been left at the set only as long as it is safe, thus if he does not move quickly he may be caught standing up at impact. This is very dangerous.

(6) The pilot will maintain intercommunication with the crew up until the last moment and warn them of the impending impact. It is not reasonable to expect the crew to remain braced for long periods. If they are not in communication with the pilot the temptation to get up and see how things are progressing may end in their being caught away from their ditching station with consequent injury. A casualty in ditching is a grave handicap to the rest of the crew, who may scarcely be able to save themselves.

m. **Drill During Ditching.**

The crew must not relax or release themselves in their ditching stations until the airplane has come to rest. The first impact of the tail can be mistaken for the shock against which they are on guard, but it will be followed by a greater shock as the nose strikes the water after a correct three pointer tail down ditching.

NOTE: Serious casualties have occurred in crews who have not taken up proper ditching stations or where they have relaxed before the final impact. Also, some crews have thought that they knew better ditching stations than those laid down in the official drill; this also has resulted in casualties. It is pointed out that these drills are the result of the experience of a great many previous ditchings and are drawn up accordingly.

n. Handling of Landplanes in Ditchings.

(1) Use of Flaps. - The flaps should be lowered to reduce the speed at which the airplane can approach and touch down. It is better to use a medium setting and not to lower them fully because little, if any, further reduction of speed is obtainable by so doing, while the rate of descent is increased and the airplane approaches more nose down. A steep nose down descent is dangerous if the sea is met sooner than expected, and also more height is required for flattening out from such an attitude.

NOTE: The maximum flap deflection for the B-25C and B-25D airplanes is 45 degrees.

(2) Approach Speed. - Assuming that symmetrical power is not available the normal glide approach speed should be used. This will insure control and some margin of speed after flattening out to allow the pilot to choose the best point for ditching on the swell.

(3) Touch Down. - Apart from choosing the best point at which to ditch, the pilot should hold off until he loses all excess speed above the stall and so strikes the sea at the normal three point landing attitude (slow landing attitudes for tricycles). The best point for ditching is towards an oncoming swell top.

(4) Direction of Approach in a Swell. - In a steep swell the pilot should ditch along the top of the swell. He should ditch up wind in a long ocean swell; however, if ditching along the swell would involve alighting with a very strong cross-wind, the airplane should be ditched into the wind. In ditching across the swell, the airplane should be put down on an upslope towards the top.

(5) Ditching Across Wind Along a Swell. - As the sea is approached drift should be taken off by side-slipping and the airplane ditched on the upslope of the swell.

(6) Use of Engines. - If one engine is available a little power should be used to flatten the approach; but the engine should not be used to such an extent that the airplane cannot be turned against it right down to the stall, with a margin of rudder power in hand. On no account should the engine be opened up during the final stages of ditching. The power that can be used will depend on the characteristics of the airplane.

o. Retention of Fuel for Ditching. - The value of power in ditching is so great that the pilot should always ditch before the fuel is quite exhausted, when it is certain that shore cannot be reached.

p. Altimeter. - The aneroid altimeter is quite unreliable as an indicator of close approach to the sea. The trailing antenna can be used, the wireless operator signalling the pilot when the current drops as the weight hits the sea. An alternative method is to engage the antenna with an insulated hook held in the hand, when the impact of the weight on the sea will be felt. This drill can only be carried out where a suitable ditching station is adjacent to the wireless operator.

q. Drill After the Airplane Has Come to Rest.

NOTE: There are two critical periods in ditching:

(1) The actual handling of the airplane on the water. This is the sole responsibility of the pilot.

(2) The abandonment of the airplane in an orderly manner after ditching in the very shortest possible time. This cannot be done well in a training fuselage in a hangar without much practice. Far less can it be expected to carry out an efficient drill in the dark after a shock in a fuselage rapidly filling with water unless the drill is perfect. Practice makes perfect. A large number of crews have saved themselves thus and been rescued by surface craft.

(a) The crew must not release themselves until the airplane comes to rest.

(b) Most multi-engine airplanes now have an automatic release for the dinghy, but do not depend solely upon this because the mechanism may have been damaged. Operate the manual release of the dinghy as soon as the airplane comes to rest but not before. The manual release should not be pulled before or during ditching to avoid inadvertent release as a result of the impact. If this mistake is made the dinghy will break out while there is still way on and it may thus drift out of reach.

(c) As soon as the airplane comes to rest, rise from the ditching stations and collect the equipment detailed in the drill. Leave by the hatch assigned in the drill and in the correct order, carrying that equipment allocated to each crew member. When the dinghy radio is carried remember that it and the means of erecting the aerial (mast or kits) are the most vital pieces of equipment required in the dinghy to assist rescue; the pigeons are next in importance and the food last but not least.

(d) On emerging inflate the life jacket if not already done. Do not be surprised to find that waves may be breaking over the airplane. If they are large it is possible to be swept off. If the airplane has a life line attached to the inside of the hatch, make use of it, otherwise hold on to the outside of the hatch and await a favorable moment to board the dinghy, but by doing so take care not to block the escape hatch or to hinder the tempo of the drill to any great extent.

(e) In airplanes equipped with blow-out dinghies one man is detailed to assist the dinghy from the stowage and it is his duty to see that the necessary cordage does not entangle during inflation. He should also assist the dingly into the water in order to hasten the boarding.

(f) If the dinghy should inflate inverted an effort should be made to right it from the wing if the airplane is not sinking too rapidly; otherwise one (and only one) of the crew should jump into the sea and right it. There are two methods of righting the dinghy:

1. If there are handling patches on the bottom of the dinghy grasp them with both hands. Then haul on these patches with the knees on the buoyancy chamber. Now while still hauling on the handling patches lean back and prepare to become submerged for a moment. Even the largest dinghy will turn over.

2. In the absence of handling patches place the toe of the foot on the bottom of the ladder, grasp the two nearest stabilizing pockets. Lean back and haul on the pockets while pressing with the foot on the ladder.

(g) Do not jump on the inverted dinghy, as doing so expels the air trapped beneath it and makes righting more difficult.

(h) If there is a painter which attaches the dinghy to the airplane, it is made light in order that it will break if the airplane sinks while the dinghy is still attached. There is a floating knife attached to the dinghy near the point where the painter is made fast. This knife is to be used to cut the dinghy free.

r. Boarding the Dinghy.

(1) If the ditching has been made into the wind the dinghy should float towards the empennage and the boarding would not be difficult.

(2) If a cross-wind ditching has been made, the airplane will tend to swing into the wind. If the dinghy is on the up wind side of the airplane, there is danger of its becoming wedged beneath the wing as the airplane rolls and swings into the wind. On the other hand, if the dinghy is on the down wind side there is a danger of its getting beneath the fuselage or empennage, which may be thrashing up and down as the airplane weather cocks into the wind. Look out for jagged edges which may puncture the dinghy.

(3) Do not jump into the dinghy; by so doing it may become damaged and the whole crew endangered.

(4) If boarding from the sea, use the rope ladder, or the tail line if provided. When using the ladder grasp the ratlines (which run across the dinghy) with one hand and the bottom rung of the ladder with the other pushing it down into the water as far as it will go to assist in inserting the foot. Then grasp the ratline with both hands and pull, at the same time pressing downward with the foot.

(5) One man already in the dinghy can be of great assistance to those in the water.

(6) To avoid the consequences of exposure it is important not to get wetter than absolutely necessary. Wet clothing must not be taken off; it is far warmer with wet clothes on than off. In hot weather this may not apply so far as cold is concerned, but the body should be covered against the sun.

(7) On every main dinghy there is a heaving line which may be used for aiding crews to reach the dinghy.

(8) All the above actions concerning the boarding of the dinghy are comparatively simple if the life jacket is fully inflated. If this jacket has been partly inflated by mouth it is important to see that the mouth valve is closed before using the CO_2 bottle. A nonswimmer can feel quite confident in a fully inflated jacket providing the leg straps are secure.

s. Aboard the Dinghy.

(1) Once aboard it is the duty of the man detailed in the drill to check whether there are any leaks and stop them up with the repair material provided. Another member of the crew is also detailed to connect up the inflating bellows and inflate until the dinghy is rigid. If any of the crew are in the water inflation of the dinghy will greatly assist in boarding.

(2) Once every one is aboard, the pilot should call the roll and give the order to cast off, then the crew should paddle away from the airplane. If the airplane floats, keep nearby to increase the chance of being spotted. But do not remain made fast to the airplane if there is any chance of the dinghy being punctured or in rough weather where the dinghy is likely to be damaged by the rise and fall of the airplane.

(3) The dinghy cover should next be rigged with the assistance of the whole crew.

(4) Once the dinghy cover is rigged, bailing should start to clear out most of the water.